Viktor Hacker, Christof Sumereder

Electrical Engineering

Also of interest

Protecting Electrical Equipment
Vladimir Gurevich, 2019
ISBN 978-3-11-063596-6, e-ISBN (PDF) 978-3-11-063928-5,
e-ISBN (EPUB) 978-3-11-063606-2

Electrochemical Energy Systems
Artur Braun, 2018
ISBN 978-3-11-056182-1, e-ISBN (PDF) 978-3-11-056183-8,
e-ISBN (EPUB) 978-3-11-056195-1

Frontiers of Science and Technology
Edited by Olfa Kanoun, 2020
ISBN 978-3-11-058407-3, e-ISBN (PDF) 978-3-11-058445-5,
e-ISBN (EPUB) 978-3-11-058414-1

Frontiers of Science and Technology
Edited by Gabriela Celani, Olfa Kanoun, 2017
ISBN 978-3-11-053623-2, e-ISBN (PDF) 978-3-11-053768-0,
e-ISBN (EPUB) 978-3-11-053626-3

Viktor Hacker,
Christof Sumereder

Electrical Engineering

Fundamentals

DE GRUYTER

Authors
Prof. Dr. Viktor Hacker
Graz University of Technology
Institute of Chemical Engineering and Environmental Technology
Inffeldgasse 25/C/II
8010 Graz
Austria
viktor.hacker@tugraz.at

Prof. Christof Sumereder
FH JOANNEUM University of Applied Sciences
Werk-VI-Strasse 46
8605 Kapfenberg
Austria
christof.sumereder@fh-joanneum.at

ISBN 978-3-11-052102-3
e-ISBN (PDF) 978-3-11-052111-5
e-ISBN (EPUB) 978-3-11-052113-9

Library of Congress Control Number: 2019951921

Bibliographic information published by the Deutsche Nationalbibliothek
The Deutsche Nationalbibliothek lists this publication in the Deutsche Nationalbibliografie; detailed bibliographic data are available on the Internet at http://dnb.dnb.de.

Cover image: Wakila / E+ / Getty Images
Typesetting: Integra Software Services Pvt. Ltd.
Printing and binding: CPI books GmbH, Leck

www.degruyter.com

We do not inherit the earth from our ancestors; we borrow it from our children.
Mahatma Gandhi

Note of thanks

The authors would like to thank all persons who contributed to the success of this book and to De Gruyter for editorial and production support. We wish to express sincere thanks to Ing. Reinhard Strasser for his invaluable help in the process of continued review and preparation of the final manuscript and illustrations and Dipl.-Ing. Andrea Jany for her assistance and attention to details throughout the process.

https://doi.org/10.1515/9783110521115-202

Preface

This textbook aims to provide access to the basics of electrical engineering for students of various disciplines. The textbook introduces the readers to the practical usages of electrical engineering and electronics. Mathematics is kept simple and only the basic relationships are listed. To gain a deeper understanding, it is recommended to work through the exercises on each topic.

Electrical engineering is a heavily interdisciplinary branch of engineering, which concerns itself with the usages of electric energy. As a sub-discipline of physics, electrical engineering, although very present in everyday life, proves to be a theoretical specialised field which is hard to grasp especially for people who are not electrical engineers. Many electrotechnical processes can only be described based on their effect, hence only indirectly.

"Current" refers to the movement of charge carriers. In the following chapters we will introduce further terms as part of a conceptual model which allows mathematically describing the movement of charge carriers and the phenomena linked to it (e.g. electrostatic repulsion or attraction). The three central terms are: "current (intensity)", "voltage" and "resistance". These three terms are repeatedly used in electrical engineering and can often be deduced from logical considerations and conclusions. In this model, the driving quantities are voltage, electric current (flow of charge carriers), and resistance that works against them.

The physical foundations of electrical processes are based on one of the four fundamental forces of physics, the electromagnetic interaction. Forces in electrical engineering are transferred through massless force carriers. In the case of electromagnetic interaction, they are called photons (light particles). Electric fields and magnetic fields are different manifestations of the same fundamental force and were jointly described as electromagnetism for the first time in the 19th century. The theoretic foundations for the correlations of the two fields are Maxwell's equations of electrodynamics. An example of direct and practical use is radio communication with satellites in space by means of electromagnetic waves that cause a change in the electric field which in turn generates a magnetic field and vice versa.

The word electricity is derived from the ancient Greek word for amber (ἤλεκτρον – *ēlektron*) and was introduced around 1600 by W. Gilbert[1] under the term "electrica". Electricity is the umbrella term in physics for all phenomena that are caused by resting or moving electric charge. Phenomena known today as electrical phenomena have been observed by mankind since early history: lightnings,

1 William Gilbert: English physician and physicist (1544–1603)

https://doi.org/10.1515/9783110521115-203

Northern Lights, St. Elmo's fire[2], attracting and repelling forces. The following list includes several historical milestones of electricity:

640 BC	The Greek philosopher and scientist Thales of Miletus observes attracting forces **when rubbing amber.**
350 BC	The **compass** is already known in China.
1200	The **compass** comes to be known in Europe.
1780	The Italian physician **Galvani** experiments with metal plates and electrolytes, inspired by frogs' legs.
1833	The French scientist Charles du Fay determines that there are two **opposed types of electric charge.**
1847	The German scientist **Werner von Siemens** invents the first **electrical cable.**
1860	Antonio Pacinotti (Italian physicist) constructs the **first direct current motor** with a multi-part commutator.
1864	The Scottish physicist James Clerk Maxwell develops the so-called **"Maxwell's Equations"** which describe how electric fields and magnetic fields are connected under given conditions.
1878	The Englishman Swan (1878) and the American Edison (1880) patent for the first viable **lightbulb.**
1889	Mikhail Dolivo-Dobrovolsky (Russian eng.) invents the **three-phase electric system** and **engine.**
1890	So-called **"War of the Currents"** between Thomas Edison (DC) and George Westinghouse (N. Tesla AC).
1895	Guglielmo Marconi (Italian entrepreneur) and Nikola Tesla develop **wireless communication.**
1904	Otto Nußbaumer first **wireless music transmission** (radio transmission)
1906	Invention of the **electron tube.**
1941	Conrad Zuse (German engineer) develops the first functioning **computer** using relays.
1948	Invention of the transistor by the Americans Bardeen and Brattein
1958	The **first integrated circuit (IC)** is developed by Jack Kilby (USA, engineer).
1960	The Austrian chemist Karl Kordesch invents the dry **alkaline battery.**
1971	The first commercially available mass-produced **microprocessor** 'Intel 4004' was put on the market.
1987	Discovery of **High Temperature Superconductivity** by Karl Alexander Müller and Georg Bednorz
1991	The first **lithium-ion battery** is available on the market.
1995	The **global navigation satellite system** is fully functional.
1999	The Personal Digital Assistant (PDA, Palm) marks the beginning of the era of **mobile devices.**
2004	First production of **Graphen** in Laboratory by Andre Geim and Konstantin Novoselov

2 St. Elmo's fire is actually continuous coronal discharge in the atmosphere. It shines in a blueish purple colour because of the spectral lines of the gases oxygen and nitrogen in the Earth's atmosphere.

Contents

1 The basic physic principles and definitions

1.1 The simple circuit

In everyday life, people do not distinguish between technically correct designations for electric quantities but abbreviate and incorrectly name it “electricity”. Colloquially, the expression “electricity bill” is used, when in reality the electrical energy consumption is meant; when an electrical accident happens, it is referred to as “electric shock”.

A person with technical knowledge is aware that a flow of an electric charge is designated “electric current” and that the physical quantity of current (intensity) uses the unit ampere. Furthermore, an expert knows that it is the voltage (measured in volts) that drives the current and that resistance (measured in ohm) at constant voltage determines the current (Figure 1.1).

Figure 1.1: Correlation between current, voltage and resistance.

To better understand the correlation between electric current, voltage and resistance, we look at the water cycle as analogue to the electric circuit.

Considering this, it becomes apparent that

- a higher voltage with constant resistance causes a higher electric current and this correlation is linear[3]: $V \sim I \quad \rightarrow I \sim V$
- a higher resistance with constant voltage causes lower electric current and the correlation is once again linear: $R \sim \frac{1}{I} \quad \rightarrow I \sim \frac{1}{R}$

Combining these two aspects directly leads to Ohm’s law: $I = \frac{V}{R}$

3 n times the voltage causes n times the current.

https://doi.org/10.1515/9783110521115-001

Table 1.1: Water cycle as analogue to electric circuit.

Water cycle (analogue)	Electric circuit
Figure 1.2: Closed water cycle.	**Figure 1.3:** Closed circuit.
The flow of water $\frac{Q}{t}$ is caused by the pressure difference ΔP generated by pump P.	The current flow is caused by the potential difference (= voltage V) generated by the voltage source.
The pressure difference ΔP determines the amount of water pumped via the load per time.	The potential difference (voltage V) determines the electric charge per time (current I) flowing through the load.
The pressure loss due to the resistance in the container C is as high as the pressure difference ΔP generated by the pump.	The voltage loss on the resistance R is as high as the generated voltage V.

1.1.1 The schematic diagram

The schematic diagram (see Figure 1.4) is a graphic representation of an electric circuit widely used in electrical engineering. It does not consider the real shape and configuration of the components but serves as an abstracted representation of the respective components' function and their electric wiring in form of standardised symbols.

In schematics, we mostly use standardised circuit symbols (see Table 1.2). These graphic symbols are defined by the IEC Standard 60617, among others, which is common in Europe. In North America, the graphic representations according to the ANSI Standard Y32 (IEEE Standard 315) are conventionally used.

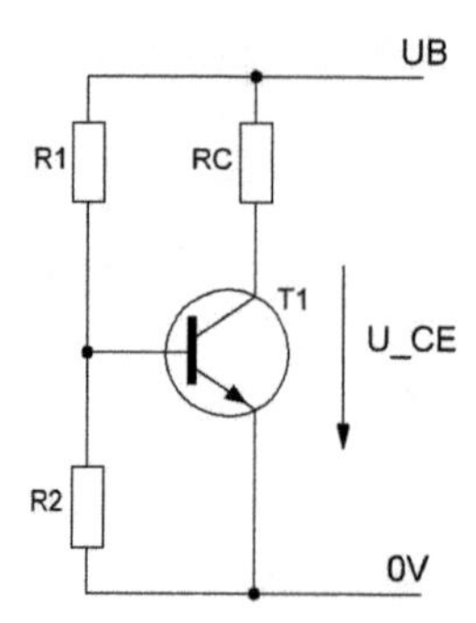

Figure 1.4: Example of a simple schematic diagram (bipolar junction transistor and 3 resistors).

An example for a circuit symbol, which is graphically represented in different ways, is the electrical resistance. According to the IEC Standard, the circuit symbol for a resistor is a rectangle, while according to the ANSI Standard it is a zigzag line.

Table 1.2: Selection of circuit symbols.

Resistor	Capacitor	Inductance	Diode	Light-emitting diode (LED)
Ideal voltage source	Ideal current source	Battery	Zener diode	Z Impedance
Transformer	Switch	Contactor or relay	Bipolar junction transistor	F Fuse
M1 M 3~Δ Motor	Optocoupler	▷ ∞ − + Op-amp	Earth	Mass, housing

1.1.2 Equivalent schematic

Equivalent schematics are used to depict electrotechnical connections in a clear way and hence support the calculability. The equivalent schematic has the same electrotechnical properties as the original circuit and consists of ideal components. The equivalent schematic of a battery in Figure 1.5 takes the real behaviour into account: the terminal voltage V_T decreases linearly with increasing load current. The equivalent schematic for a conductive wire becomes an ohmic resistor if direct current passes through it (see Figure 1.6). Convoluted resistor networks can be redrawn as a clear equivalent schematic with the same electrical properties (see Figure 1.7).

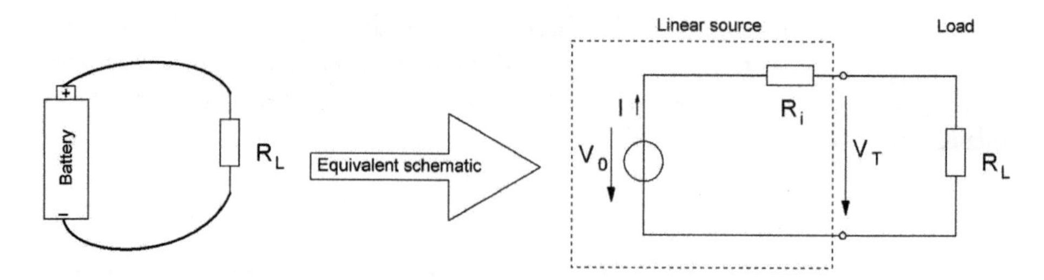

Figure 1.5: Equivalent schematic of a real voltage source.

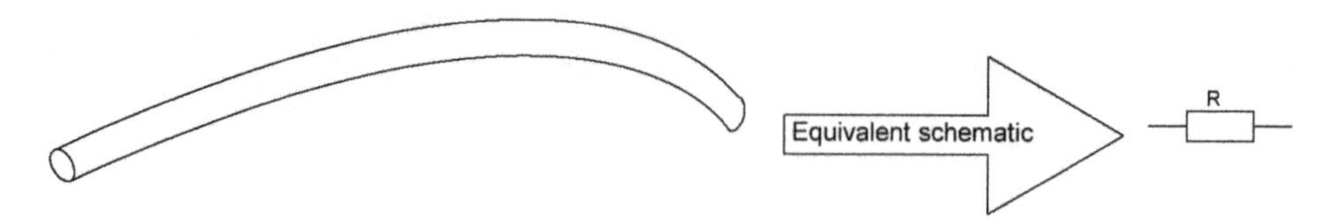

Figure 1.6: Equivalent schematic of an electric conductor.

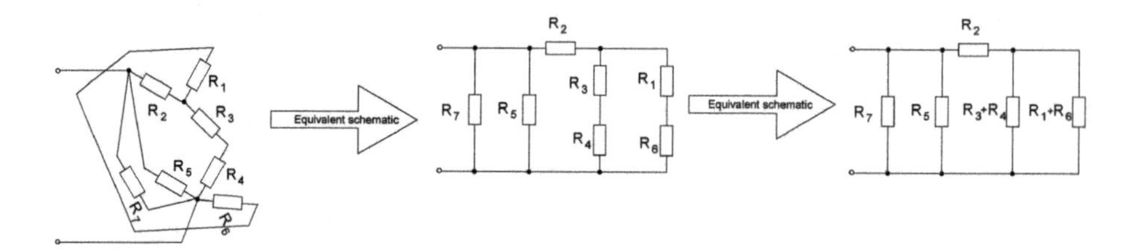

Figure 1.7: Equivalent schematic of a resistor network.

1.2 Electric charge Q

The essence of electricity lies in the existence, the build-up and equalisation of charges. Charge is a property linked to matter that has the characteristic of a quantity (therefore "quantity of electric charge") and manifests itself in forces[4] between charges.

$$Q = N \cdot e$$

Q Charge in As
N Number of particles
e Elementary charge in As

4 This is described by Coulomb's law.

The charge Q consists of countable elementary charges.[5] Their carriers are parts of the atoms and molecules (mobile or as space charge).

1.3 Current I

The correlation between current and electric charge, by definition, is as follows:

$$I = \frac{Q}{t}$$

I Current in A
Q Charge in As
t Time in s

The current I is divided by an (arbitrarily oriented) area. Q is the overall charge that flows through in the time t. If the charge transfer is viewed within an infinitesimal time interval dt, it is written as:

$$I = \frac{dQ}{dt}$$

The defining equation for the charge Q results from this relation:

$$Q = \int I \cdot dt$$

or for stationary[6] current flow:

$$Q = I \cdot t$$

1.4 Mechanisms of the electric current flow

Electric processes are based on the existence of electric charges that can move freely. According to the Bohr model[7] each atom consists of a nucleus and negatively charged electrons that circle the nucleus. In turn, the nucleus consists of positively charged protons and uncharged neutrons. In general, an atom has as many electrons as protons (it is electrically neutral). Protons, electrons and charged atoms are „**charge carriers**". Atoms with a different number of protons and electrons are called **ions**.

5 The charge is "quantized".
6 time-independent.
7 The Bohr model was developed in 1913 by the Danish physicist Niels Bohr and can be compared to the planetary system. The electrons move around the nucleus in elliptical orbits.

Quantity of electric charge

The electron carries the smallest possible negative charge e^-; the proton carries the same positive elementary charge e^+. The elementary charge is a fundamental physical constant:

$$e^- = -1,6 \cdot 10^{-19} C$$

$$e^+ = +1,6 \cdot 10^{-19} C$$

In practice this unit is too small. The unit coulomb C is used. The following applies:

$$1\ C = 6,24 \cdot 10^{18}\ e$$

The electric charge Q can be calculated from the number of electrons and protons. The electrons circle the nucleus only in specific orbits. Those electrons with roughly the same distance are compiled into a group, called electron shell. The electrons in the (usually not completely full) outer shell are called valence electrons. The electrons in the shells below are called core electrons.

Free electrons

Electrons usually are bound to their atom. In some substances, especially in metals, atoms are arranged in a way that some electrons pass from the outer shell into the sphere of neighbouring atoms. These electrons do not directly belong to an atom anymore – they become free electrons. Generally, we can say that there is approx. one free electron in each metal atom. The entirety of free electrons causes metals to be electrically conductive.

1.4.1 Electric conduction in metals

Metals are structured like a space lattice. The nodes of this lattice are the positively charged atomic nuclei that are bound to their spot in the lattice. In between there are free electrons, which can be viewed as an electron gas. Free electrons move under the influence of external forces (e.g. an electric field caused by an applied voltage) and thereby cause an electron flow. The electron flow always generates magnetism. Due to the friction of the electrons within the atomic lattice, losses occur, and heat is generated. These two effects can be observed as a common interaction in practical electrical engineering as for example in induction machines (e.g. transformer and motor) or in heating wires.

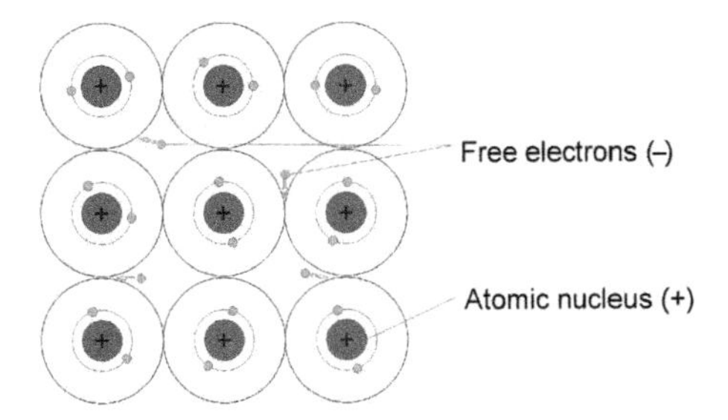

Figure 1.8: Free electrons and positive atomic nuclei of a conductor.

1.4.2 Electric conduction in liquids

The conduction in liquids is also based on the transport of electrically charged particles. However, these particles are not electrons, as in solids, but positively and negatively charged ions.

Example 1.1: Two electrodes made of platinum or nickel are submerged into a container filled with distilled water (see Figure 1.9) and connected to a direct voltage source V. With distilled water (a) almost no current flows, because there are hardly any charge carriers in the liquid. If we carefully add little amounts of acid, alkaline solution or salt (b), then the current increases many times over. These substances create mobile charge carriers that get accelerated by the voltage source (through the emerging electric field).

Chemical compounds that are dissociated[8] in solid, liquid or gaseous state and exhibit a directed movement under the influence of an electric field are called electrolytes.

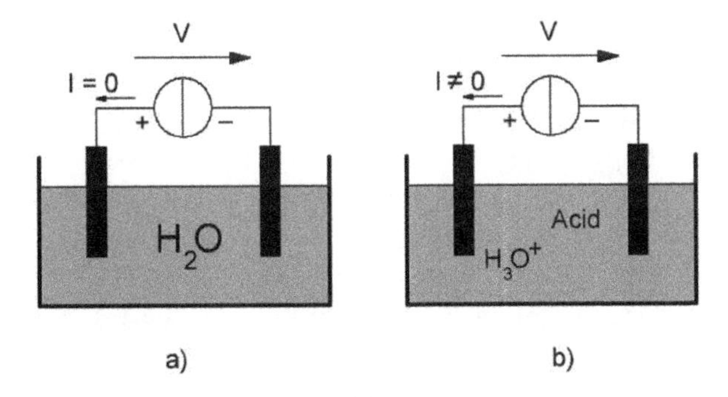

Figure 1.9: Electric conduction in liquids.

8 In chemistry, dissociation means the stimulated or autonomous division of chemical compounds into two or more molecules, atoms or ions.

More detailed explanation with a watery $CuSO_4$ solution: The dissociating force triggered by the strong dipolar character of the H_2O molecule causes $CuSO_4$ to split into positive cations (Cu^{2+}) and negative anions ($SO_4{}^{2-}$). Copper lacks 2 electrons and SO_4 has 2 excess electrons. In a $CuSO_4$ solution the following processes take place:

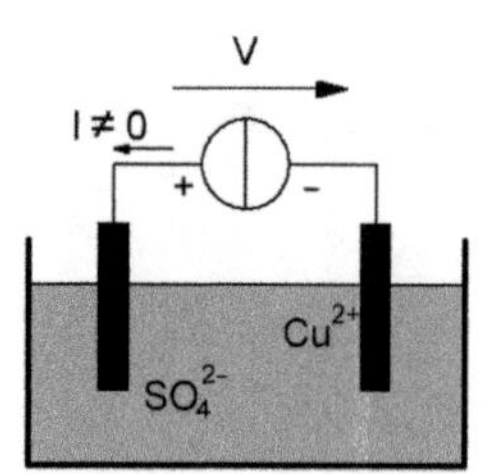

Figure 1.10: $CuSO_4$ solution.

Due to an applied direct voltage and the resulting electric field the Cu^{2+} ions advance in the direction of the field towards the negative electrode (=cathode). There they gain electrons, are neutralised and deposited in form of high-purity metals. If the anode is made from copper, the $SO_4{}^{2-}$ residue reacts to $CuSO_4$; a copper atom goes into solution. The copper anode is decomposed; the concentration of the copper sulphate solution is retained.

1.4.3 Electric conduction in gases

Gases and vapours in their normal state and at low field strengths are complete nonconductors. However, if we ionise gas molecules (through removing or adding one or more electrons) they follow the force of the electric field. The minimum amount of energy required to remove an electron is called ionisation energy (unit eV – electron-volt).

Ionisation can occur through:

- Electromagnetic radiation (light, X-radiation, nuclear radiation, cosmic radiation)
- Impact ionisation through moving electrons

Impact ionisation is significant for technical applications: If an electron is ejected by another electron, the gas molecule turns into a positive ion that moves towards the cathode on the electric field, where it induces an electron and neutralises itself. If a gas particle catches an electron, it is negatively charged and moves towards the anode on the electric field.

The current (=mobile electrons) continuously produces new charge carriers. The concentration of said charge carriers is essential for the current. The existing voltage does not influence them (other correlations than Ohm's law). It may even happen that the necessary voltage decreases with increasing current. An ionisation can also be caused by impacts of positive ions. However, this does not occur as often as the development of carriers through electron impacts.

When ionising, there are constantly new impacts. The mean path that an electron covers between two impacts is called **mean free path** and is a measure for the energy stored during the acceleration. Besides the type of gas, the mean free path depends on the pressure. It is shorter for ions than for electrons. As a result of continuous impacts of the free electrons and positive ions with the gas molecules, the created carriers transform. During this process, primarily negative ions are produced by the deposition of electrons on atoms or molecules.

Consequently, there are three types of charge carriers in the gas:
- (negative) electrons
- positive ions
- negative ions

The continuous new formation of charge carriers ought to lead to an increase in carrier density. It appears nevertheless that – by recombining charge carriers – a state of equilibrium is reached (positive and negative charge carriers mutually neutralise themselves).

We differentiate between
- **self-sustaining** and
- **dependent gas discharge.**

In a **dependent gas discharge**, ionisation does not produce enough charge carriers to maintain current. Therefore, this process does not play an essential role, but can be seen as preliminary stage for a self-sustaining gas discharge.[9]

After applying a sufficient voltage at the electrodes, a **self-sustaining gas discharge** occurs and the current increases. The discharge also persists without outer ionisation. The self-sustaining gas discharge produces its charge carriers through impact ionisation of gas atoms and ejection of electrons at the negative electrode when positive ions occur.

1.4.4 Electric conduction in a vacuum

An absolute vacuum, i.e. the complete absence of atoms and molecules is practically unobtainable. Nowadays, pressures of up to $10^{-6}\ mbar$ $(0.1 \cdot 10^{-6}\ kPa = 10^{-4}\frac{N}{m^2})$ can be reached with diffusion pumps. This corresponds to approx. 10^{10} molecules per cm^3.

9 This effect is used when measuring the intensity of X-rays.

The conduction mechanism in the vacuum is maintained by free electrons. There is only electronic conduction. To induce or discharge electrons, two electrodes consisting of metal parts and located in an evacuated glass container are used.

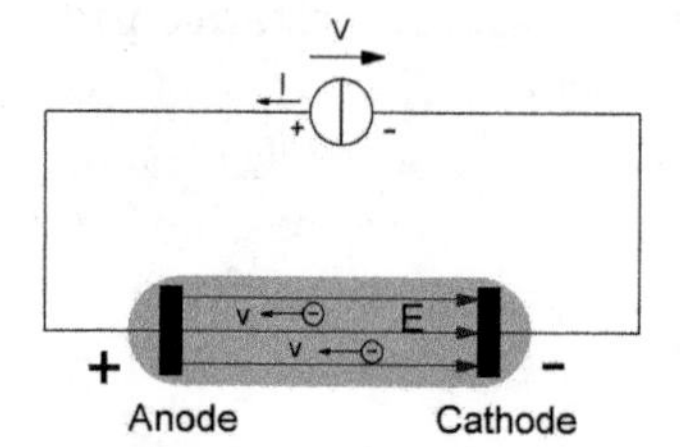

Figure 1.11: Vacuum tube.

If there is an electric voltage between anode and cathode, the electric field affects the electrons. This electric field defines the force F towards the anode. The cathode always provides new electrons, therefore the constant current I, also known as electron beam, flows. The force that is exerted on an electron is proportional to the electric field strength E.

$$F = e \cdot E$$

$$E = 10^8 \frac{V}{m}$$

$$F = 1,6022 \cdot 10^{-11} N$$

This force in itself is very small. Considering that an electron has the extremely little mass $m_e = 9,1 \cdot 10^{-31}$kg, the electron is strongly accelerated with $a = \frac{F}{me} = 0,175 \cdot 10^{20} \frac{m}{s}$. The great mean free path results in a steady acceleration with the velocity v:

$$v = a \cdot t$$

$$s = \frac{1}{2} a \cdot t^2$$

$$s = \frac{1}{2} \cdot (v^2/a) \cdot t^2$$

$$\mathrm{v} = \sqrt{2 \cdot \mathrm{a} \cdot \mathrm{s}}$$

After a distance of 1 mm the velocity of the electron amounts to:

$$v = \sqrt{2 \cdot 0,175 \cdot 10^{20} \cdot 1 \cdot 10^{-3}} = \underline{1,87 \cdot 10^8 \frac{m}{s}}$$

That is approximately half of the speed of light. If the electron speed enters a range near the speed of light, according to the theory of relativity, the mass of the electron increases compared to the rest mass:

$$m = \frac{m_0}{\sqrt{1 - \left(\frac{v}{c}\right)^2}}$$

m_0 Rest mass of the electron in kg
c Speed of light in m/s $\left(c \approx 3 \cdot 10^8 \frac{m}{s}\right)$
v Velocity in m/s

If the electron e^- is moved a certain distance by the force, work $W = e \cdot V$ is executed and the kinetic energy increases. Due to conservation of energy (work is transformed into kinetic energy), one can set up the following equation

$$e \cdot V = \frac{m \cdot v^2}{2} \Rightarrow v = \sqrt{\underbrace{\frac{2 \cdot E}{m}}_{\substack{factor \\ for\, v \ll c\; const.}} \cdot V}$$

The traversed voltage is therefore a measure for the kinetic energy of an electron. The unit electron-volt (eV) is used for electron energy $\left(1\ eV = 1,6022 \cdot 10^{-19}\ Ws\right)$.

To remove the electrons from the cathode a specific form of work, the work function, has to be performed. It lies in the range of a few electron-volts.

Electrons are bound to nuclei by electric forces. The removal from the metal surface (emission) may occur via:

a) **Field emission**: Sufficiently high field strength dissolves electrons out of the metal compound.
b) **Photoemission**: Light particles (photons) of high energy eject electrons.[10]
c) **Thermionic emission**
d) **Secondary effects:** e.g. collision of positive ions onto the metal surface.

1.5 Current direction

The direction of conventional current is defined, independent of the type of charge carriers, as the flow of charges from the positive pole to the negative pole. This corresponds to the flow of positive charge carriers, from a given (higher) potential to a lower potential.

10 The physicist Albert Einstein explained this with the "photoelectric effect", for which he received the Nobel Prize for Physics.

The actual movement of negatively charged electrons in electrical circuits is opposite to the direction of the defined conventional current.

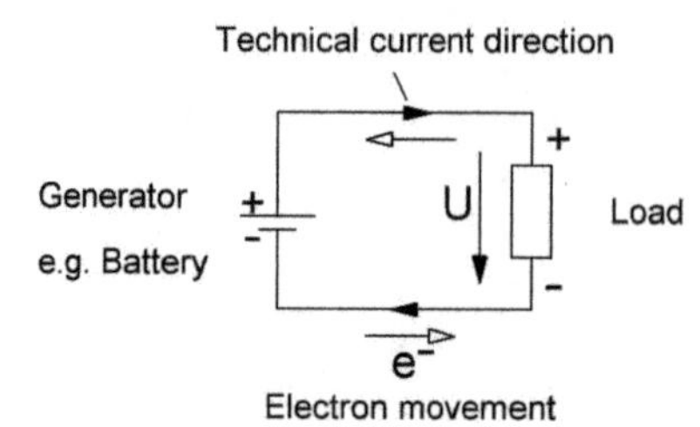

Figure 1.12: Technical current direction vs. actual movement of electrons.

Historical determination of the conventional current direction

One of the first electrical phenomena investigated was the deposition of metals by electric current. The positive current direction was observed by the flow of the positive metal ions in the electrolyte from the positive to the negative electrode. Electrons were not yet known. This definition has preserved until today.

Example 1.2 Generally, we get negatively charged anions (e.g. acid residue SO_4^{2-}, CO_3^{2-}, hydroxide OH^-) at the positive anode[11] and positively charged cations (e.g. hydrogen H^+, ammonium NH_4^+, metals like Cu^{2+}, Fe^{2+}, Ag^+, Al^{3+}) at the negatively charged cathode.

The reactivity of the alkali metals increases with the atomic mass (caesium would be the most reactive alkali metal). The alkali metals react with water and build a single charged cation, while the water is decomposed to hydroxide ions and hydrogen gas.

For example: Chloralkali-electrolysis

When dissolving sodium chloride in water – ions are formed through dissociation. If the saturated, aqueous solution is electrolyzed (graphite electrodes), chloride ions are oxidized to chlorine at the anode. At the cathode, water decomposes into hydrogen and hydroxide ions. The negative charged hydroxide ions react with the positively charged sodium ions and form sodium hydroxide (NaOH).

11 Per definition, the anode is the electrode where oxidation reactions (electron release into the circuit) happen. The cathode is the electrode where reduction reactions (electron induction of electrons that are supplied by the circuit) like metal deposition or H_2 development takes place. In high voltage applications the cathode is the electrode where electrons were set free to gas phase.

1.6 Alternating and direct current

Direct current (DC) is characterized by the unidirectional flow of electric charge, or a system in which the movement of electric charge is in one direction only.

When the current changes repeatedly the direction with a definite frequency, the current is known as alternating current (AC). The usual waveform is a sine wave. If the direct current is superimposed on an alternating current, this is called universal current (see Table 1.3).

Table 1.3: Current types.

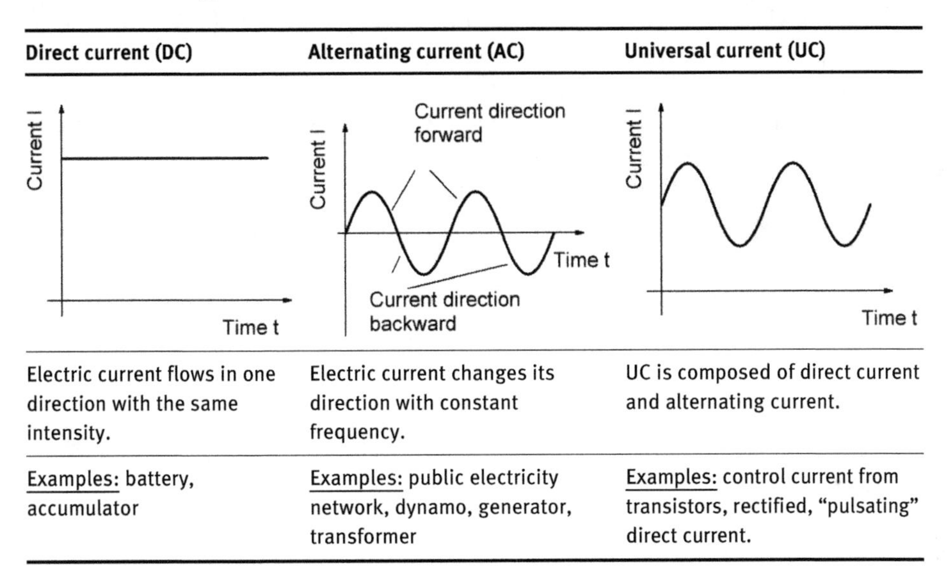

Direct current (DC)	Alternating current (AC)	Universal current (UC)
Electric current flows in one direction with the same intensity.	Electric current changes its direction with constant frequency.	UC is composed of direct current and alternating current.
Examples: battery, accumulator	Examples: public electricity network, dynamo, generator, transformer	Examples: control current from transistors, rectified, "pulsating" direct current.

1.7 The electric field

Each **electric charge** puts the surrounding space into a special state. In this region, electric charges experience a force.[12] The special state is called **electric field.**

Electric fields are caused by electric charges and temporal changes in magnetic fields. If the electric field is created by a stationary electric charge, we talk about an **electrostatic field.**

Each electric charge is surrounded by an electric field. Electric charges interact and affect each other:
- Like charges repel each other.
- Unlike charges attract each other.

12 The force can be calculated by means of Coulomb's law.

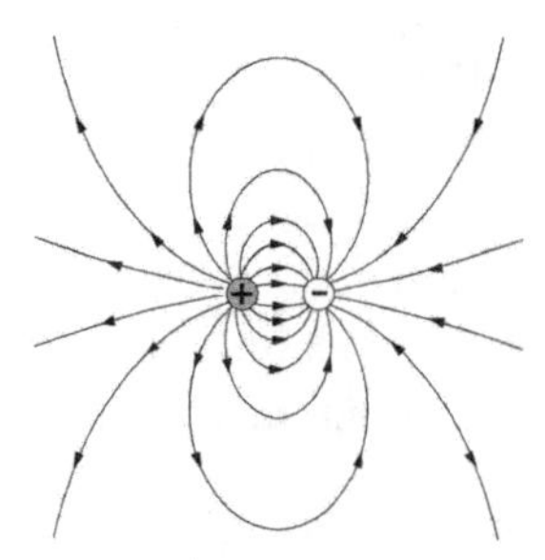

Figure 1.13: Electric field of a dipole.

A charged particle moves along an imaginary line in the electric field. Such lines of occurring forces are called electric field lines.

The following applies for electric field lines:

- **Direction:** The electric field lines start at the **positive charge (source)** and end at the **negative charge (sink).**
- Electric field lines always start and end **on the surface** of an electric charge.
- Electric field lines always leave the surface of the electric charge **in a right angle.**
- Field lines **do not intersect.**
- The **denser the field lines,** the stronger the force exerted on the charges.

From a mathematical point of view, the electric field is a vector field of the electric field strength. It assigns a vector for **direction and magnitude** of **electric field strength** to each point in the space.

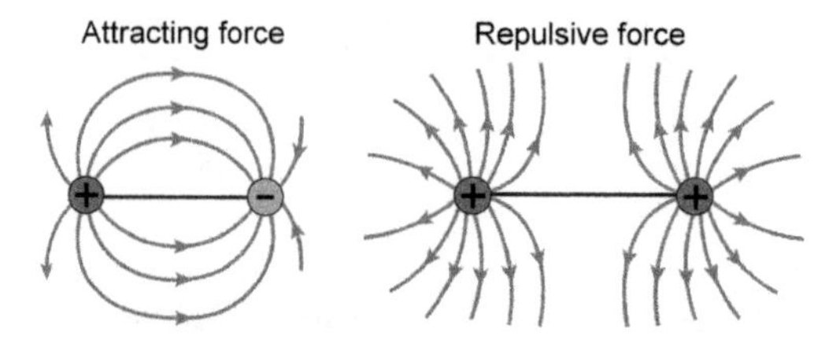

Figure 1.14: Attracting force and Repulsive force of charges.

The electric field is the cause for forces exerted on charges. According to the first of Newton's axioms, a body accelerates as long as force is exerted on it. Moving charge carriers are called current. The sources of an electric field are positive charges, sinks and negative charges. The electric field [Vm^{-1}; NC^{-1}] is uniform when neither its magnitude nor its direction changes from one point to another.

1.7.1 Force on charged particle in electric field

The electric field strength is the measure of the force, a charged body experiences in an electric field. The existing electric field strength at any point in the field can be defined as the force exerted on a positive point unit charge $Q_+ = 1\ As$ located there.

Defining equation:

$$\vec{F} = E \cdot q$$

The force causes an acceleration of the charged particle with the mass m

$$\vec{a} = \frac{\vec{F}}{m}$$

1.7.2 Force between point charges

Coulomb's law describes the force F between two point charges or spherically distributed charges Q_1 and Q_2 within a medium. The attracting or repelling forces F of individual charges are very small. They were measured for the first time by the French physicist Coulomb using a highly sensitive torsion balance. The spherical distribution of field lines around charges results in **Coulomb's law:**

$$F = \underbrace{\frac{1}{4 \cdot \pi \cdot \varepsilon}}_{\text{proportionality – constant}} \cdot \frac{Q_1 \cdot O_2}{r^2}$$

with $\varepsilon = \varepsilon_0 \cdot \varepsilon_r$

F	Force in N
Q_1, Q_2	Charge in As
ε_0	Absolute dielectric constant of the vacuum $\varepsilon_0 = 8.859 \cdot 10^{-12} \frac{F}{m}$
ε_r	Relative dielectric constant[13] (indicates the factor of deviation of the dielectricity from the vacuum) for air is $\varepsilon_r = 1.00059$
r	Distance between the charges in m

13 The relative dielectric constant is often referred to as "relative permittivity".

Table 1.4: Examples of ε_r.

Medium	ε_r
Vacuum	1 (by definition)
Air	1.00059
Org. insulators	2 to 4
Glasses	5 to 10
Water	81

1.8 Electric potential φ

Electric potential φ is a scalar quantity equal to the amount of work needed to move a unit charge from a reference point (e.g. earth) to a specific point inside the electric field. In Figure 1.15 two horizontally charged plates and one positively charged object on the lower plate are shown. The potential energy of the object on the lower level 1 is $W_{p,1}$ and on the upper level 2 $W_{p,2}$. If the object is moved from level 1 to level 2 by means of the acting force $F_A = -F_{el}$ while overcoming the electric force $F_{el} = Q \cdot E$, according to the general definition of work $W = \int \vec{F} \cdot d\vec{s}$, the lifting of the object equals the work for the vertical lifting along a field line opposite to the field direction. Then, the following applies

$$W_{1,2} = \int_1^2 \overrightarrow{F_A} \cdot d\vec{s} = -\int_1^2 \overrightarrow{F_{el}} \cdot d\vec{s}$$

$$W_{1,2} = -\int_1^2 \overrightarrow{F_{el}} \cdot d\vec{s} = -\int_1^2 \underbrace{\left|\overrightarrow{F_{el}}\right|}_{Q \cdot E} \cdot |d\vec{s}| \cdot \underbrace{\cos\left(\overrightarrow{F_{el}}, d\vec{s}\right)}_{-1} = \int_1^2 Q \cdot E \cdot ds = QEl$$

This work equals the difference in potential energy of the object from level 2 and level 1:

$$W_{1,2} = W_{p,2} - W_{p,1}$$

The potential energy of the object on level 2 equals the energy on level 2 plus the work needed for the transport from 1 to 2:

$$W_{p,2} = \underbrace{\int_1^2 \overrightarrow{F_A} \cdot d\vec{s}}_{\substack{\text{work needed} \\ \text{for transport}}} + W_{p,1}$$

$$W_{p,2} = -Q \cdot \int_1^2 \vec{E} \cdot d\vec{s} + W_{p,1} = QEl + W_{p,1}$$

In order to obtain a characteristic value for the field, the potential energy refers to the unit charge, i.e. W_p is considered an object of the charge $Q_+ = 1\ As$ at a certain point of the field and this value is called the **potential φ** of the field at that point.

Defining equation:

$$\varphi = \frac{W_p}{Q_+}$$

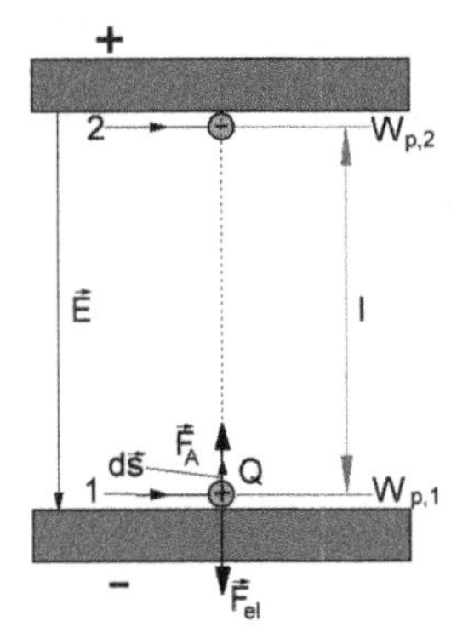

Figure 1.15: Derivation of the correlation between field strength and potential.

1.9 Voltage *V*

Voltage is the difference in electric potential between two points. In a static electric field, voltage is defined as the work needed to move a unit charge between two points.

The necessary charge to transport a charge Q_+ from object 1 to object 2 equals the difference of the potential energies $W_{1,2} = W_{p,2} - W_{p,1}$ of the charges on the two objects.

The potential difference equals the work for the unit charge $Q_+ = 1\ As$:

$$\varphi_2 - \varphi_1 = \frac{W_{1,2}}{Q}$$

It is called **voltage *V*** ([14]).

14 In German-language regions, the symbol U is used for voltage.

Defining equation: $V = \frac{W_{1,2}}{Q_+}$

$[V] = V\ (Volt)$[15] SI unit: $[V] = \frac{J}{As}$

The voltage between two objects therefore is the work that is necessary to transport the unit charge $Q_+ = 1$ As from the negatively charged object to the positively charged object. If this voltage is expended, the charges are separated.

Voltage can cause electric current: If there is a voltage between two particles and these two particles are connected through a conductor, current flows. (More specifically: There is an electric field between the two particles. This field exerts force on the free charge carriers, which causes them to move.)

Voltage is connected to force: Two particles, with an electric voltage between them, carry a different number of positive and negative electric charges, respectively; therefore, they are positively and negatively charged relative to one another. The respective surplus charges exert forces of attraction on each other.

Technical generation of voltage:

- Electric charges are separated by **induction** (generator, transformer) due to magnetic energy.
- A chemical reaction separates electric charges in the **galvanic cell.**
- Heat separates electric charges in the **thermocouple.**
- Radiant energy of the light separates electric charges in the **photovoltaic cell** (industrially in photovoltaic panels).
- Pressure separates electric charges in some **piezoelectric crystals** (e.g. quartz).[16]

1.10 Ohm's law

Ohm's law describes a linear correlation in certain electric conductors between voltage drop V and the electric current I which is flowing through the conductors at constant temperature:

$$\boldsymbol{V = R \cdot I}$$

V Voltage in V
R Electrical resistance in Ω (Ohm)
I Electric current in A

15 The unit for voltage was named after the Italian physicist Alessandro Volta (1745–1827).

16 The occurrence of voltage on solids, when they are elastically deformed, is called (direct) piezoelectric effect. Reversely, materials are deformed when a voltage is applied (converse piezoelectric effect).

The resistance R ideally[17] is constant, regardless of the magnitude of the current. If this is the case, we talk about **linear circuits.**

1.10.1 Resistance R and conductance G

The resistance R of a conductor depends on the material and the spatial dimensions of the conductor. To measure the resistivity of the material, the specific resistance is defined as the resistance of a cube with edges 1 m in length.

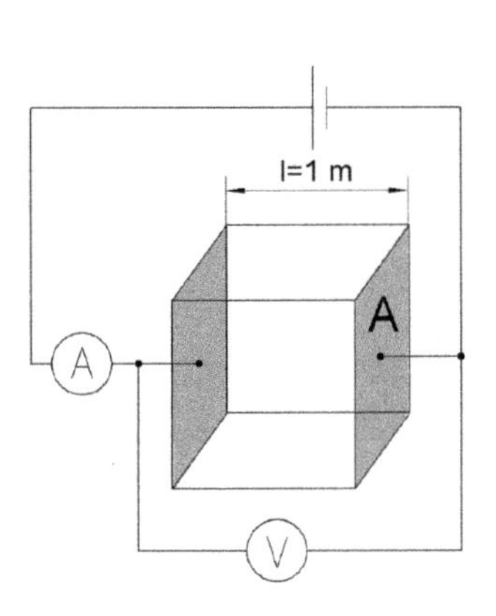

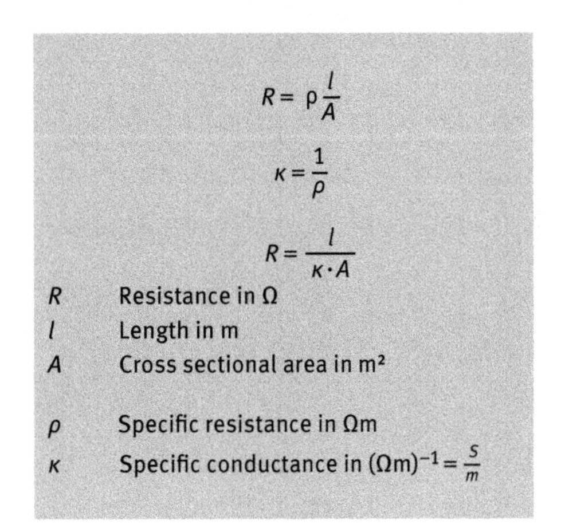

Figure 1.16: Resistance R.

The unit of resistance is a quantity derived from I and V. A conductor has a resistance of 1Ω when a voltage $V = 1\ V$ drives a current of $I = 1\ A$ through the conductor.[18]

The technically used resistors can generally be divided into two groups: wire resistors and sheet resistors. The materials of the wire filaments or coatings, usually attached to cylindrical ceramic bodies, have specific resistances with low temperature dependence. The resistance is determined by measuring current and voltage.

Instead of stating the resistance of an object, sometimes the reciprocal dimension, the **conductance G** is given (e.g. to compare the conductivity of different conductors).

---

17 In reality, there is at least a temperature dependency.

18 For example, a mercury column with a length of l=1.063 m and a cross-section of A=1 mm² has a resistance of 1 Ω. Such an Hg column is used as a standard for the laboratory.

$$G = \frac{1}{R}$$

G Electrical conductance in S (Siemens)

1.10.2 Temp. dependence of R, resistivity ϱ, conductivity σ, temperature coefficient α

The resistance of conductor materials changes with the temperature (e.g. increases for metals). This change in resistance is non-linear and the accurate characteristic curve can be approximated using a polynomial. The resistance R_2 of a conductor at the temperature ϑ_2 can be calculated if its resistance R_1 is known at the temperature ϑ_1:

$$R_2 = R_1 \cdot [1 + \underbrace{\alpha_1}_{\substack{linear\\TC}} (\vartheta_2 - \vartheta_1) + \underbrace{\beta_1}_{\substack{quadratic\\TC}} (\vartheta_2 - \vartheta_1)^2 + \gamma_1(\vartheta_2 - \vartheta_1)^3 \ldots]$$

Up to $\vartheta_2 = 100$ °C, generally, only the linear temperature coefficient α_1 is used.

$$R_2 = R_1 \cdot [1 + \underbrace{\alpha_1(\vartheta_2 - \vartheta_1)}_{vital\ term\ for\ change}\ ]$$

For most pure metals the linear temperature coefficient α_{20} is around $4 \cdot 10^{-3}\ C^{-1}$ and the quadratic temperature coefficient β_{20} is around $0.6 \cdot 10^{-6}\ C^{-1}$. Pure metals have the highest temperature coefficients and therefore are not suitable as standard or measuring resistor. They can, however, be used as measured-value transmitter (resistance thermometer).

Through alloys, the temperature coefficients can be reduced in such a way that temperature dependence practically does not exist anymore. These materials are used for standard and measuring resistors (e.g. constantan, manganin).

Table 1.5: Conductivity, resistivity (for 25°C), temperature coefficient at 20°C.

Material	Conductivity σ in MS/m	Resistivity ρ in Ωm	Resistivity ρ in $\Omega mm^2/m$	α_{20} in $\frac{1}{K}$
Silver	61.7	$0.0162 \cdot 10^{-6}$	0.0162	$3.7 \cdot 10^{-3}$
Copper	58.0	$0.0172 \cdot 10^{-6}$	0.0172	$3.9 \cdot 10^{-3}$
Aluminium	36.6	$0.0273 \cdot 10^{-6}$	0.0273	$3.7 \cdot 10^{-3}$
Brass	14.2–33	$0.0704 \cdot 10^{-6}$ $- 0.0303 \cdot 10^{-6}$	0.0704 − 0.0303	$1.6 \cdot 10^{-3}$
Manganin	2.3	$0.435 \cdot 10^{-6}$	0.435	$0.01 \cdot 10^{-3}$
Constantan	2	$0.5 \cdot 10^{-6}$	0.5	$5 \cdot 10^{-5}$
Iron	9.6	$0.0104 \cdot 10^{-6}$	0.104	$5.6 \cdot 10^{-3}$
Gold	45	$0.0222 \cdot 10^{-6}$	0.0222	$5.6 \cdot 10^{-3}$

Superconductor

Superconductors are materials with an electrical resistance that abruptly equals zero if the transition temperature drops below a certain point. Many metals, but also other materials, become superconductive below their transition temperature – better known as "critical temperature" T_C. For most materials this temperature is very low. To achieve superconductivity, the material generally has to be cooled with liquid helium (boiling temperature −269 °C). Cooling with liquid nitrogen (boiling temperature −196 °C) is only sufficient with high temperature superconductor. To achieve the superconducting condition, beside the critical temperature also the critical current density J and the critical magnetic field strength B have to be fulfilled.

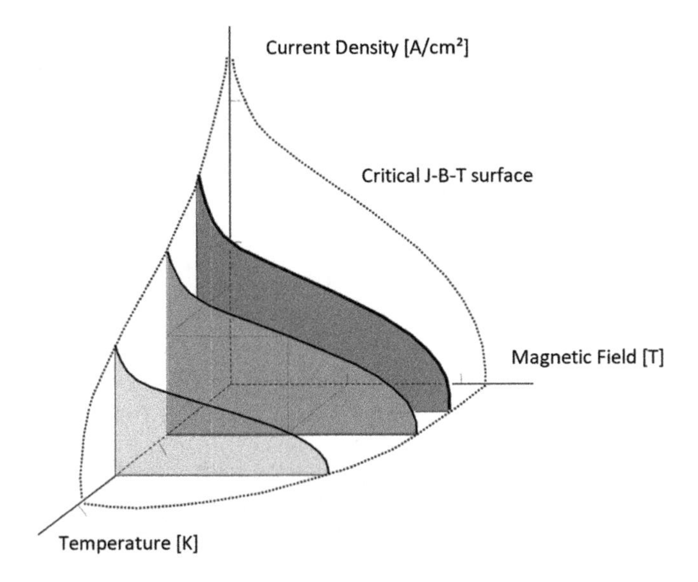

Figure 1.17: Superconducting surface.

1.11 Energy and electrical power

Electrical energy is provided by the combination of electric current and electric potential in an electrical circuit. The mechanical work $W = \int \vec{F} \cdot d\vec{s}$ is comparable to electrical work done on a charged particle by an electric field.

If the force (F) is used to lift an object by the distance (s), mechanical work is carried out. The object now has higher energy content by that amount (potential energy). This energy can perform work by e.g. letting the object drop.

Electric energy the following applies:

$$W = Q \cdot V = V \cdot I \cdot t$$

V Voltage in V
I Current in A
t Time in s
Q Electric charge in As
W Electrical work/energy in Ws

The electric charge represents the product of current I multiplied by time ($Q = I \cdot t$). Therefore, the following applies:

$$W = \underbrace{V \cdot I}_{P} \cdot t = P \cdot t$$

P Power in W

The work performed per time unit is called power P.

$$P = \frac{W}{t}$$

With direct current, the electric power P that is transformed in an electric load is the result of the voltage V multiplied by the current I:

$$P = V \cdot I$$

$$[P] = W$$

Power is one of the most important parameters for electrical machines and devices. By insertion of the Ohm's law, the equation can be transformed to:

$$\mathbf{P} = \underbrace{I \cdot R}_{V} \cdot I = \mathbf{I^2 \cdot R}$$

or also:

$$P = \frac{V^2}{R}$$

Power hyperbola

All V-I pairs of values that lead to the same power, e.g. $P = 2\ W$, result in a power hyperbola in the graphic representation. Therewith, the permissible voltages or flows of the resistors can be calculated with a given permissible power. At every point the hyperbola is calculated (for the 2 Watt example):

$$P = V \cdot I = 2\ W$$

If the ohmic resistance curves are drawn into the same diagram, the curves intersect the power hyperbola at the points with the highest permissible voltage and the highest permissible current, respectively (see Figure 1.18).

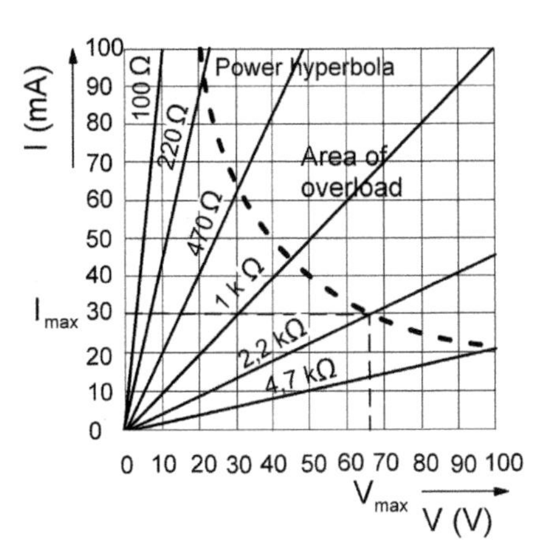

Figure 1.18: Power hyperbola.

Example 1.3

a) Based on the figure to the left, determine the values V_{max} and I_{max} for a resistance of 2.2 *kW*, 2 watt.

b) Check the values using mathematical calculations

SOLUTION

a) From Figure 1.18: $V_{max} \approx 66\ V$ und $I_{max} \approx 30\ mA$

b) $U_{max} = \sqrt{P \cdot R} = \sqrt{2 \cdot 2200} = \mathbf{66.3\ V}$

$$I_{max} = \sqrt{\frac{P}{R}} = \sqrt{\frac{2}{2200}} = \mathbf{0.03\ A}$$

1.12 Review questions

?

1) How does the simplest way of a circuit look like?
2) How does the conduction work in metals?
3) How is the technical current flow defined?
4) How is the electric field defined?
5) Which characteristics are important for a direct and an alternating quantity?
6) How do you recognise the presence of an electric field?
7) What is the difference between the terms potential and voltage?
8) When does the ohmic law apply? Does it apply to non-linear resistors?
9) Do conductive materials (such as copper) have a negative or positive temperature coefficient?
10) What are the conductivities of copper, silver and gold?
11) Take a look at the difference between power and work. Give examples

1.13 Exercises

EXERCISE 1.1

A lightning stroke has a charge Q of 2 C during a period t of 50 µs. Which current I is present in average?

How long gives a flashlight light when there is a lamp with a rated current I of 250 mA inside?

EXERCISE 1.2

Which ohmic resistance R has a round copper conductor with a cross-section dimension A of $4mm^2$ and a length l of 1km?

EXERCISE 1.3

A lamp of a flashlight has a rated Voltage V of 9 V and a rated Current of 0.5A. Which Energy W is consumed during 1 h?

EXERCISE 1.4

An electric heater consumes a current I of 11 A at a line voltage of 230 V. Which electric power P does the heater consume?

2 Electrical networks

2.1 Kirchhoff's circuit laws

2.1.1 Kirchhoff's current law (KCL) or Kirchhoff's first law

Currents flowing into the node are counted positively; currents flowing out of the node are counted negatively. The law is also called nodal rule. At any node, the sum of currents flowing into that node I_{in} is equal to the sum of currents flowing out of that node I_{out} or equivalently:

The sum of currents at a nodal point is zero.

$$\sum I = 0$$

$$\sum I_{in} = \sum I_{out}$$

$$I_1 + I_2 = I_3 + I_4 + I_5$$

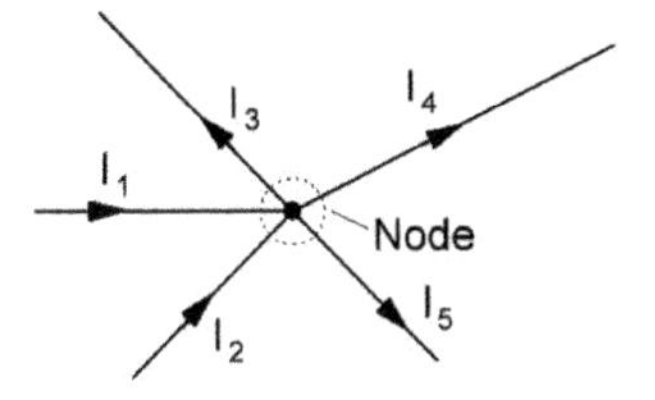

Figure 2.1: Kirchhoff's current law.

2.1.2 Kirchhoff's voltage law (KVL) or Kirchhoff's second law

After a complete cycle in a mesh, the voltage has returned to its starting potential. Consequently, the sum of all voltages in a mesh at any instantaneous value is zero. Voltages that flow in the direction of the mesh are counted positively; voltages that flow against the direction of the mesh are counted negatively. The law is also called mesh rule:

https://doi.org/10.1515/9783110521115-002

The sum of the voltages (potential differences) in a mesh is zero.

$$\sum_{i}^{n} V_i = 0$$

$$V_{q1} - V_4 - V_3 - V_{q2} + V_2 + V_1 = 0$$

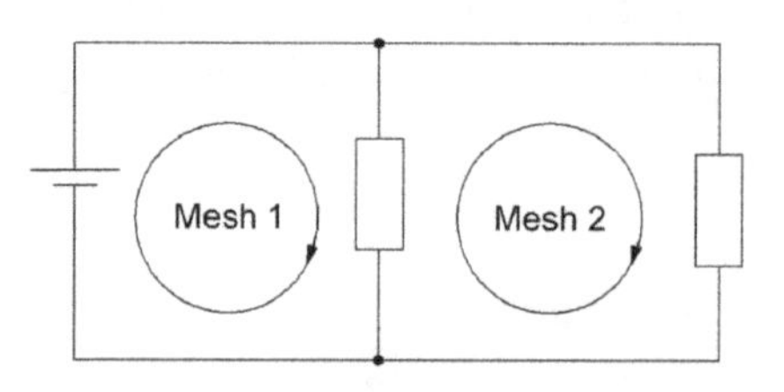

Figure 2.2: Network with two meshes.

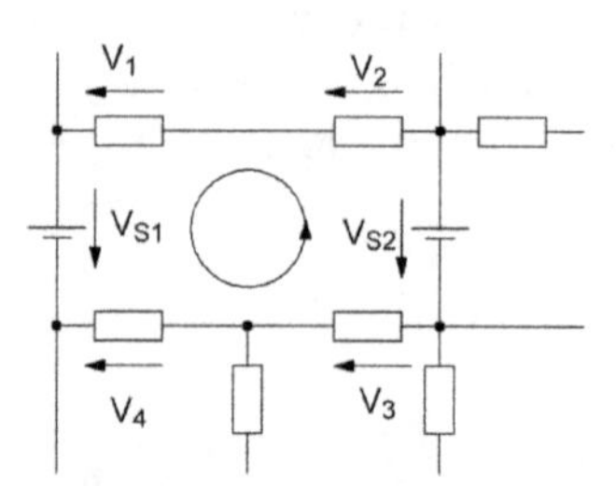

Figure 2.3: Network with ohmic resistors and voltage sources.

2.2 Series connection of resistors

For the series connection of n resistors, the following applies:

$$R_t = \sum_{1}^{n} R_n = R_1 + R_2 + \ldots + R_n$$

2.3 Parallel connection of resistors

For the parallel connection of n resistors, the following applies:

$$\frac{1}{R_t} = \frac{1}{R_1} + \frac{1}{R_1} + \ldots + \frac{1}{R_n}$$

or written with the conductance:

$$G_t = G_1 + G_2 + \ \dots \ + G_n$$
$$R_t = \frac{1}{G_t}$$

Total conductance = $\sum$ individual conductance

R_t in a parallel connection is always smaller than the smallest individual resistance.

For the parallel connection of n resistors with the same resistor value the total resistance $\frac{1}{n}\cdot$ individual resistance.

For **the parallel connection of 2 resistors** the following applies:

$$R_t = \frac{1}{\frac{1}{R_1} + \frac{1}{R_2}} = \frac{R_1 \cdot R_2}{R_1 + R_2}$$

In parallel connections with 2 resistors of the same resistor value, the total resistance amounts to half the individual resistance.

2.4 Combining linear and non-linear resistors

2.4.1 Series connection of linear and non-linear resistance

Example 2.1: Resistor and diode in forward direction

Find the total resistance.

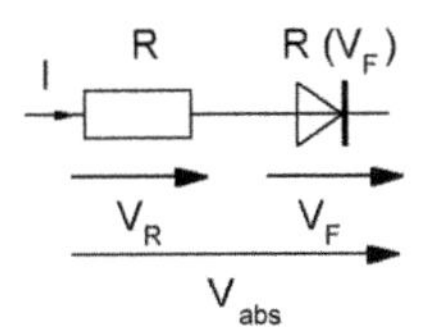

Figure 2.4: Series connection of resistor and diode.

Solution: Mathematical or graphical

A mathematical solution proves difficult as the exact diode characteristic curve is difficult to determine. Therefore, an empirically determined typical characteristic, which is indicated graphically in data sheets, is commonly used. The graphical solution is, hence, feasible.

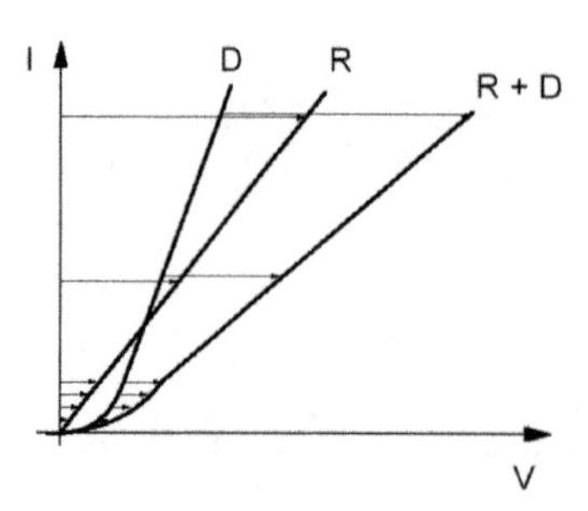

Figure 2.5: Solution principle – series connection resistance and diode.

Solution principle: Addition of the partial voltages at constant current

Results in one point respectively on the resulting total characteristic curve R + D (see Figure 2.5).

2.4.2 Parallel connection of linear and non-linear resistors

Example 2.2: Transistor input (base) with parallel input divider resistor

Find $R_t = R_{B2}||r_{BE}$

Again, the graphical solution is more feasible.

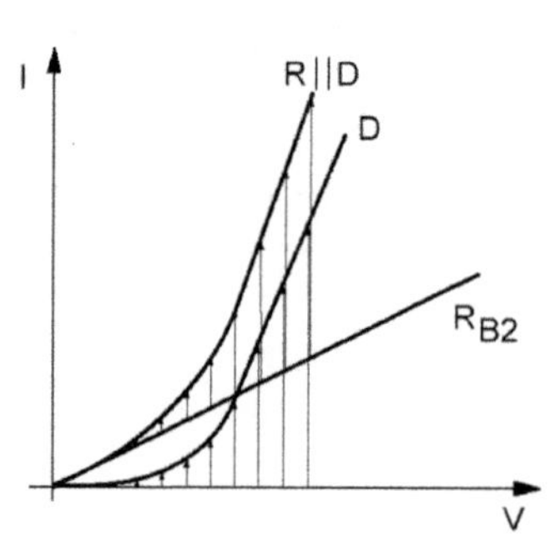

Figure 2.6: Solution principle – parallel connection of resistance and diode.

Solution principle: The partial currents at respective constant voltage are added up.

The sum results in one point on the new and shared resistance characteristic curve $R||D$.

2.5 Wye (Y) connection and delta (Δ) connection

Wye (also star connection) and delta connections are mainly used in three-phase AC technology.

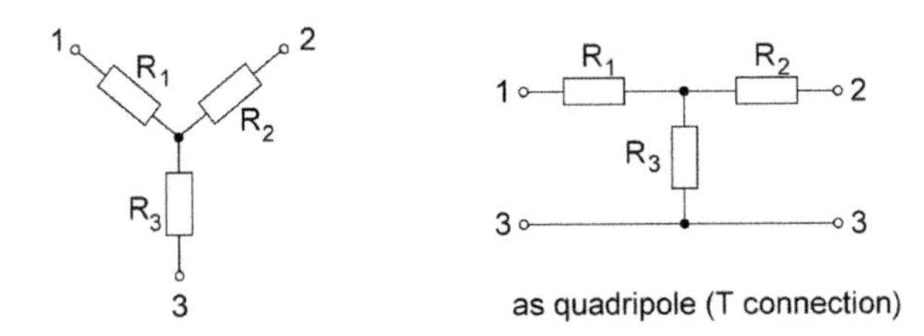

Figure 2.7: Wye connection.

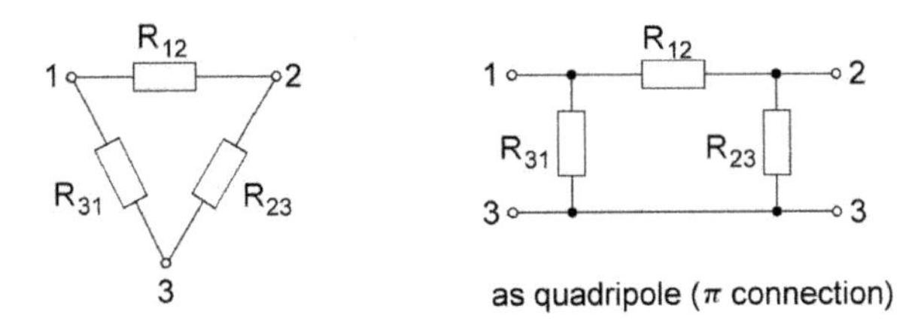

Figure 2.8: Delta connection.

2.5.1 Conversion wye connection → delta connection

Secondary conditions when determining the conversion formulas: The resistance between 2 points must be equal in both circuits.

Point	Y	Δ
1 – 2	$R_1 + R_2 = R_{12}\|\|(R_{31} + R_{23}) =$	$\frac{R_{12} \cdot (R_{31} + R_{23})}{R_{12} + R_{23} + R_{31}} = R_{E12}$
2 – 3	$R_2 + R_3 = R_{23}\|\|(R_{12} + R_{31}) =$	$\frac{R_{23} \cdot (R_{12} + R_{31})}{R_{12} + R_{23} + R_{31}} = R_{E23}$
3 – 1	$R_3 + R_1 = R_{31}\|\|(R_{23} + R_{12}) =$	$\frac{R_{31} \cdot (R_{23} + R_{12})}{R_{12} + R_{23} + R_{31}} = R_{E31}$

If R_{12}, R_{23}, and R_{31} are given, R_1, R_2 and R_3 can be calculated:

$$R_1 = R_{E12} - R_2 \quad R_2 = R_{E23} - R_3 \quad R_3 = R_{E31} - R_1$$

$$R_1 = R_{E12} - R_{E23} + R_{E31} - R_1$$

$$2 \cdot R_1 = R_{E12} - R_{E23} + R_{E31}$$

$$R_1 = \frac{1}{2} \cdot (R_{E12} - R_{E23} + R_{E31})$$

$$R_1 = \frac{R_{12} \cdot R_{31} + R_{12} \cdot R_{23} - R_{23} \cdot R_{31} - R_{23} \cdot R_{12} + R_{31} \cdot R_{12} + R_{31} \cdot R_{23}}{2 \cdot (R_{12} + R_{23} + R_{31})} = \frac{2 \cdot (R_{12} \cdot R_{31})}{2 \cdot (R_{12} + R_{23} + R_{31})}$$

$$R_1 = \frac{R_{12} \cdot R_{31}}{R_{12} + R_{23} + R_{31}}$$

Through cyclic interchange of the indices we derive:

$$R_1 = \frac{R_{12} \cdot R_{31}}{R_{12} + R_{23} + R_{31}} \quad (1) \qquad R_2 = \frac{R_{23} \cdot R_{12}}{R_{12} + R_{23} + R_{31}} \quad (2) \qquad R_3 = \frac{R_{31} \cdot R_{23}}{R_{12} + R_{23} + R_{31}} \quad (3)$$

2.5.2 Conversion delta connection → wye connection

R_1, R_2 and R_3 are given and R_{12}, R_{23} and R_{31} can be calculated.

Approach: Ratio formation with (1), (2) and (3)

$$\frac{R_1}{R_2} = \frac{R_{31}}{R_{23}} \quad \frac{R_2}{R_3} = \frac{R_{12}}{R_{31}} \quad \frac{R_3}{R_1} = \frac{R_{23}}{R_{12}}$$

(1) exemplary divided by R_{31}

$$R_1 = \frac{R_{12}}{\frac{R_{12}}{R_{31}} + \frac{R_{23}}{R_{31}} + 1} = \frac{R_{12}}{\frac{R_2}{R_3} + \frac{R_2}{R_1} + 1}$$

solve with respect to R_{12}

$$R_{12} = R_1 + R_2 + \frac{R_1 \cdot R_2}{R_3}$$

with cyclic interchange, this results in

$$R_{12} = R_1 + R_2 + \frac{R_1 \cdot R_2}{R_3} \qquad R_{23} = R_2 + R_3 + \frac{R_2 \cdot R_3}{R_1} \qquad R_{31} = R_3 + R_1 + \frac{R_3 \cdot R_1}{R_2}$$

2.5.3 Delta-to-wye & wye-to-delta conversion in mixed circuits

Generally, individual resistances can be added up gradually (parallel or series connection) to a resistance R.

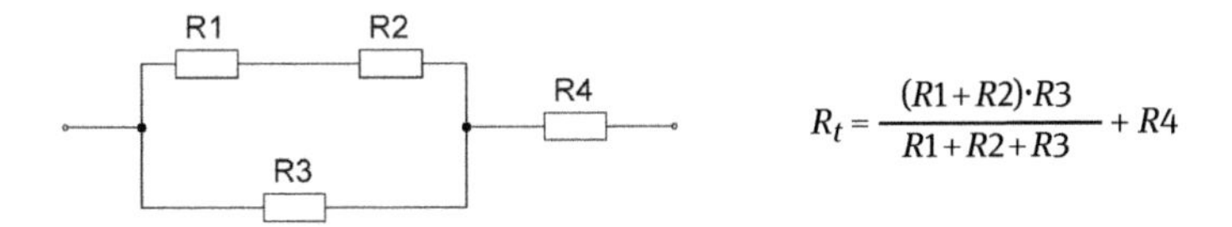

Figure 2.9: Delta-to wye conversion – initial situation.

Sometimes, however, the summation of resistances is not possible. To solve this problem, we use the wye-to-delta conversion or the delta-to-wye conversion.

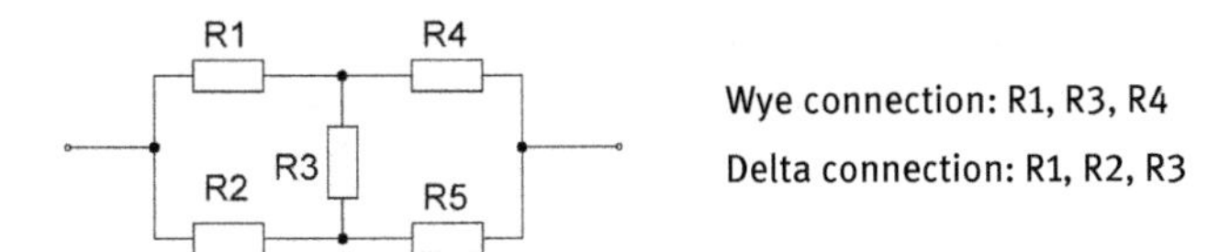

Figure 2.10: Delta-to-wye conversion – after conversion.

2.6 Mixed circuits

Mixed circuits consist of a minimum of three components.

- Combination of mixed circuits: network
- Determining the equivalent resistance
- The circuit is solved from the inside out
- Summation of series or parallel connection to equivalent resistance
- Repeat steps

A general network consists of branches, nodes and meshes.

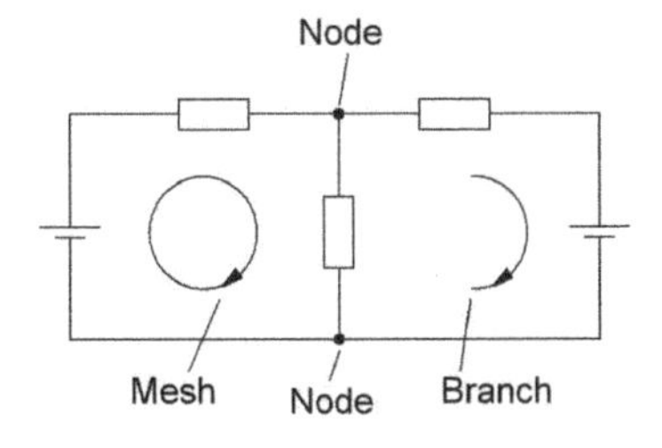

Branch: chain of two-terminal networks within a network; passed through by the same current

Nodes: points of connection between two or more branches

Mesh: closed chain of branches; a closed circuit

Figure 2.11: Mesh, node and branch.

Example of a general network

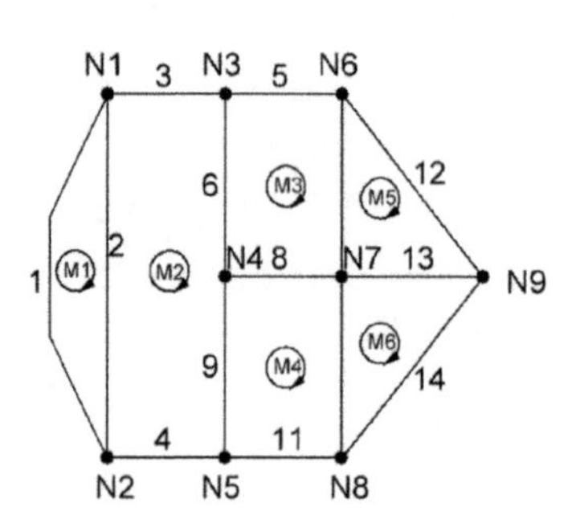

Figure 2.12: General network.

Number of branches b 14

(meaning 14 unknown currents)

Number of meshes m 6

(establish using the mesh rule)

Number of nodes n 9

(establish using the nodal rule)

$$b = m + (n - 1)$$

b = 6 + (9 − 1) = 14 equations

linear system of equations with b unknowns and b (independent) equations; solvable

2.7 Sources

2.7.1 Ideal voltage source / real voltage source

Ideal voltage source

The ideal voltage source (see Figure 2.13) maintains a constant voltage V_s independent from the load resistance between its terminals. The internal resistance of the source equals zero. Therefore, an ideal voltage source can emit any amount of power. The load current I_R is defined only by the load resistance R.

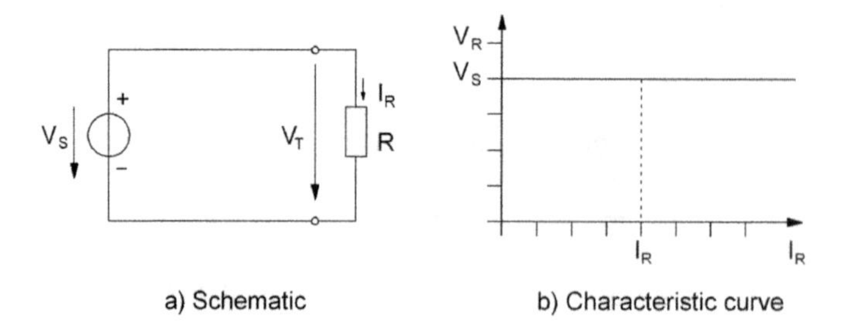

Figure 2.13: Ideal voltage source.

$$V_T = V_s = constant$$

$$V_{Ri} = 0$$

$$I_R = \frac{V_s}{R}$$

Real voltage source

A real voltage source cannot generate any amount of power. It can be observed that the terminal voltage V_T decreases with increasing load current. If the source is short-circuited, the maximum current that can be generated by the source I_K flows. A real voltage source acts like an ideal voltage source with an **internal resistance** $\boldsymbol{R_i}$ greater than zero.

$$V_{Ri} > 0$$

$$V_T = V_s - V_{Ri} \cdot I_R$$

A real voltage source can be characterised by the source voltage V_s, which in open circuit is equal to the terminal voltage, and by the short-circuit current I_{sc}:

Open circuit voltage	$I_R = 0;$	$V_S = V_t$
Short circuit	$V_t = 0;$	$I_{SC} = \frac{V_S}{V_{Ri}}$

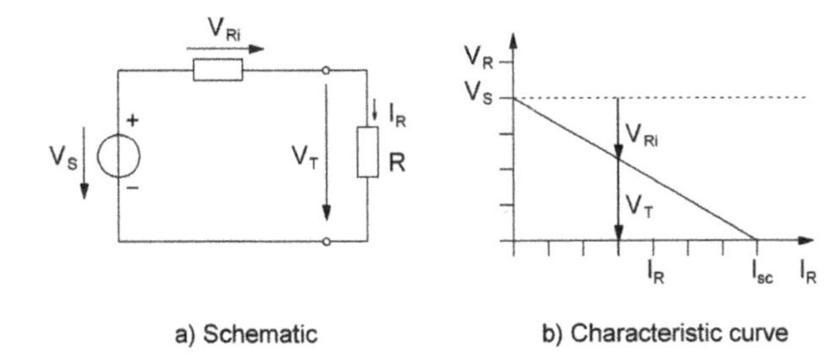

Figure 2.14: Real voltage source.

2.7.2 Ideal current source / real current source

Ideal current source

The ideal current source generates a constant current I_s independent from the load resistance. The terminal voltage is determined only by the load resistance R.

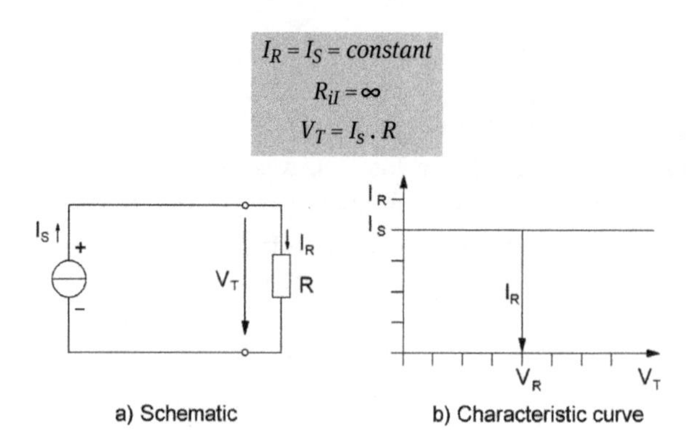

Figure 2.15: Ideal current source.

Real current source

A real current source cannot generate any amount of power. It can be observed that the load current I_R decreases with increasing resistance R. In open circuit (without load resistance or with an infinitely high load resistance) the open circuit voltage $V_T = V_0$ occurs at the terminals (terminal voltage). That is the maximum voltage which can be provided by the source. A real current source acts like an ideal current source with the internal resistance $R_i < \infty$.

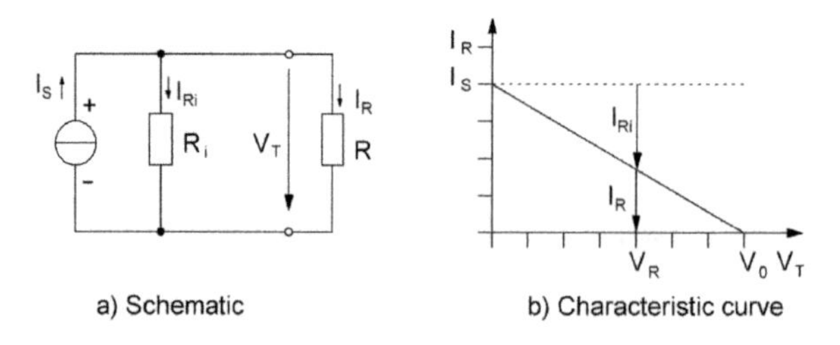

Figure 2.16: Real current source.

$$I_T = I_S - I_{Ri} = I_S - \frac{V_T}{R_{iI}}$$

$$R_{iI} > 0$$

A real current source is characterised by the source current $I_s = I_T$ and the open circuit voltage V_0

Short circuit $V_T = 0$ $I_K = I_S$

Open circuit voltage $I_R = 0$ $V_T = I_S \cdot R_i$

2.7.3 Power adjustment

The power supplied to the load resistance R by a real voltage source can be calculated as follows:

$$P_R = V_R \cdot I_R = \frac{V_R^2}{R} = I_R^2 \cdot R$$

$$V_R = V_s \cdot \frac{R}{R + R_i}$$

$$I_R = I_s \cdot \frac{R_i}{R + R_i}$$

$$P_R = V_s^2 \cdot \frac{R}{(R + R_i^2)}$$

$$P_R = I_s^2 \cdot R_i^2 \frac{R}{(R + R_i)^2}$$

The maximum power can be transmitted if the function $P_R(R)$ has a maximum. Therefore, the first derivative of the function with respect to R must be set equal to zero.

$$P_R = max \Rightarrow \frac{dP_R}{dR} \underset{=}{\text{def}} 0$$

$$0 = \frac{2(R + R_i)R - (R + R_i)^2}{(R - R_i)^4}$$

$$R = R_i$$

The maximum power can be transmitted to the load **if the load resistance R equals the internal resistance R_i of the source.** This applies to voltage sources as well as current sources due to the interchangeability of sources.

2.8 Voltage divider

In electronics, often voltages of different levels are required. For circuits with low power consumption (e.g. for amplifier inputs) this requirement is easily achieved by using a voltage divider. A series connection with two resistors constitutes the voltage divider (see Figure 2.17). If the voltage is to be continuously adjustable, a variable resistor (potentiometer) can be used instead (see Figure 2.18). If a load is connected to the voltage divider, a current I_L flows. This is called a loaded voltage divider (see Figure 2.19).

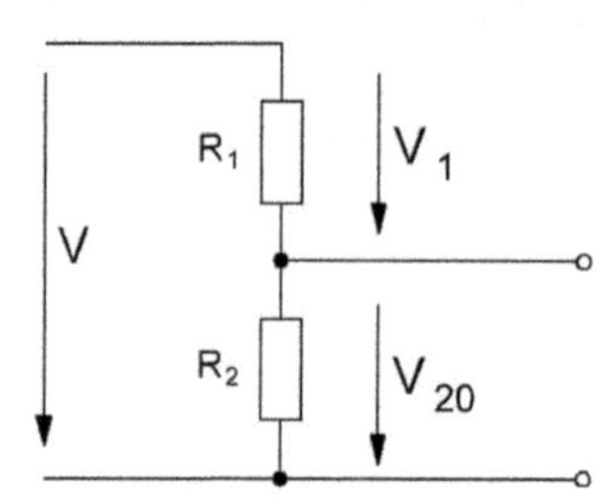

Figure 2.17: Unloaded voltage divider.

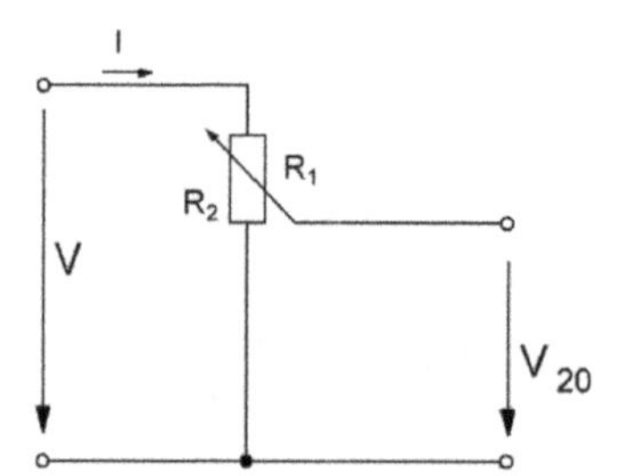

Figure 2.18: Voltage divider with potentiometer.

Voltage divider rule:
The voltages are related to each other in the same way as the resistors to which the voltages are connected.

Considering the unloaded voltage dividers in Figure 2.17 and Figure 2.18, the following applies:

$$\frac{V_{20}}{V} = \frac{R_2}{R_1 + R_2} \rightarrow V_{20} = \frac{R_2}{R_1 + R_2} \cdot V$$

The following applies for a **loaded voltage divider**:

$$V_2 = f(R_L)$$

$$V_2 = V \cdot \frac{R_{2L}}{R_1 + R_{2L}} \quad \text{with } R_{2L} = \frac{R_2 \cdot R_L}{R_2 + R_L}$$

$$q := \frac{I_c}{I_L} = \frac{R_L}{R_2}$$

q Ratio of cross current

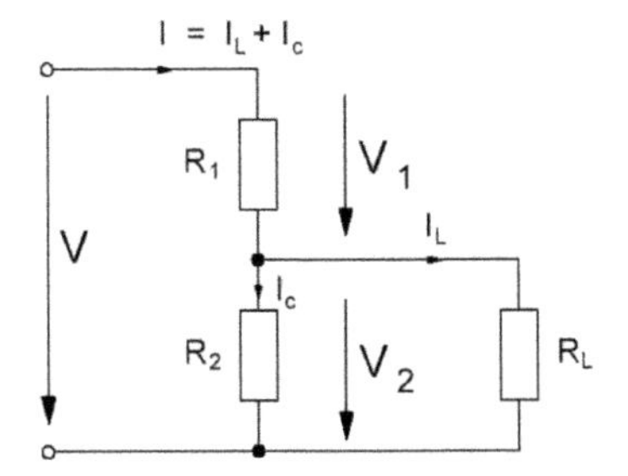

Figure 2.19: Loaded voltage divider.

I_c Cross current in A
I_L Load current in A

The output voltage V_2 of the loaded voltage divider is all the more stable, the higher the cross current I_c compared to the load current I_L. The values for the ratio of cross current q from 2 to 10 depend on the stability requirement.

2.9 The complex calculation in electrical engineering

2.9.1 Definitions

We know from mathematics, that complex numbers can be represented in a complex number plane (also referred to as "Gauss number plane"). In electrical engineering, decisive parameters can be represented as vectors. It is hence advisable to use complex calculation for calculations in electrical engineering. The advantage is that additions/subtractions of vectors can be traced to the additions/subtractions of complex numbers. A multiplication or division can be performed in a simpler way using complex notation with the corresponding calculation rules. While in mathematics, the letter i is used for the imaginary unit, in electrical engineering one uses the letter j as the letter i already represents electric current, which could lead to confusion.

Graphic forms

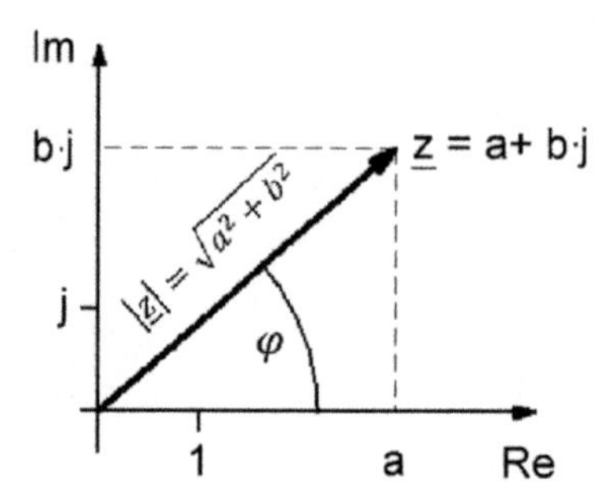

Figure 2.20: Complex number plane.

A complex quantity $\underline{z}$ is hence represented by

$$\underline{z} = a + bj$$

with the real part $\mathrm{Re}\{\underline{z}\} = a$ and $Im\{\underline{z}\} = b$ and the imaginary unit j. To indicate that the quantity is complex, the quantity symbol is underlined (or in bold type in printed works). This notation is called "Cartesian notation".[19]

For multiplication and division, the exponential form is of importance:[20]

$$\underline{z} = r \cdot e^{j\varphi}$$

A complex number in Cartesian notation can be converted into exponential form and vice versa. For the conversion, the following applies:

$$a = r \cdot \cos(\varphi) \quad b = r \cdot \sin(\varphi) \quad r = \sqrt{a^2 + b^2} \quad \varphi = \arctan \tfrac{b}{a}$$

Complex conjugates

For every complex number $\underline{z}$ there is a number $\underline{z}^*$ which is its mirror image on the real axis; the two numbers are complex conjugates of each other and differ only in the algebraic sign of their imaginary part.

$$\underline{z} = a + b \cdot j \Rightarrow \underline{z}^* = a - b \cdot j$$

19 It is also called "algebraic notation".

20 It is also called "Euler's notation".

Calculation rules

Given are the complex numbers $\underline{z_1}$ and $\underline{z_2}$:

$$\underline{z_1} = a_1 + b_1 \cdot j = r_1 \cdot e^{j\varphi_1} \qquad \underline{z_2} = a_2 + b_2 \cdot j = r_1 \cdot e^{j\varphi_2}$$

Addition and subtraction

The addition of both numbers $\underline{z_1}$ and $\underline{z_2}$ is given by:

$$\underline{z_1} + \underline{z_2} = (a_1 + a_2) + (b_1 + b_2) \cdot j = a_3 + b_3 \cdot j = r_3 \cdot e^{j\varphi_3}$$

The subtraction of both numbers $\underline{z_1}$ and $\underline{z_2}$ is given by:

$$\underline{z_1} - \underline{z_2} = (a_1 - a_2) + (b_1 - b_2) \cdot j = a_3 - b_3 \cdot j = r_3 \cdot e^{j\varphi_3}$$

When adding/subtracting, the respective real parts and the respective imaginary parts are added/subtracted.

Multiplication and division

The multiplication of the numbers $\underline{z_1}$ and $\underline{z_2}$ is given by

$$\underline{z_1} \cdot \underline{z_2} = (a_1 \cdot a_2 - b_1 \cdot b_2) + (a_1 \cdot b_2 + a_2 \cdot b_1) \cdot j = r_1 \cdot r_2 \cdot \underbrace{e^{j\varphi_1} \cdot e^{j\varphi_2}}_{e^{j(\varphi_1 + \varphi_2)}} = \underbrace{r_3 \cdot e^{j\varphi_3}}_{\substack{r_3 := r_1 \cdot r_2 \\ \varphi_3 := \varphi_1 + \varphi_2}}$$

The division of the numbers $\underline{z_1}$ and $\underline{z_2}$ is given by (expand with conjugated complex denominator in order to make it real):

$$\underline{z_4} = \frac{\underline{z_1}}{\underline{z_2}} = \frac{a_1 + j \cdot b_1}{a_2 + j \cdot b_2} = \frac{(a_1 + j \cdot b_1) \cdot (a_2 - j \cdot b_2)}{(a_2 + j \cdot b_2) \cdot (a_1 - j \cdot b_1)} = \frac{a_1 a_2 + b_1 b_2 + j \cdot (-a_1 b_2 + a_2 b_1)}{a_2^2 + b_2^2}$$

$$\underline{z_4} = \frac{\underline{z_1}}{\underline{z_2}} = \frac{r_1 \cdot e^{j\varphi_1}}{r_2 \cdot e^{j\varphi_2}} = \frac{r_1}{r_2} \cdot e^{j(\varphi_1 - \varphi_2)}$$

As additions and subtraction can be performed without difficulties, it is recommended to do so in Cartesian notation. The multiplication or division should be performed in exponential form. Converting one form into the other should be performed by means of a calculator/computer in order to benefit from the time advantage.

2.9.2 Application of the complex calculation in AC calculation

The requirements for using complex calculations when performing AC calculations include that currents and voltages can only produce sinusoidal quantities and that the components R, L, and C in the network are linear.

The voltage in an arbitrary phase position φ is given by:

$$v(t) = \hat{v} \cdot \sin(\omega t + \varphi_v)$$

The current in an arbitrary phase position φ is given by:

$$i(t) = \hat{i} \cdot \sin(\omega t + \varphi_i)$$

A complex voltage $\underline{v}$ is characterised by:
- Amplitude $\hat{v}$
- Phase position φ_V
- Angular frequency $\omega = 2 \cdot \pi \cdot f$

$$\underline{R} = R = R \cdot \underbrace{e^{j \cdot 0^\circ}}_{1} \quad \underline{X_C} = \frac{1}{j \cdot \omega \cdot C} = \frac{1}{\omega \cdot C} \cdot e^{j \cdot (-90^\circ)} \quad \underline{X_L} = j \cdot \omega \cdot L = \omega \cdot L \cdot e^{j \cdot 90^\circ}$$

Time representation

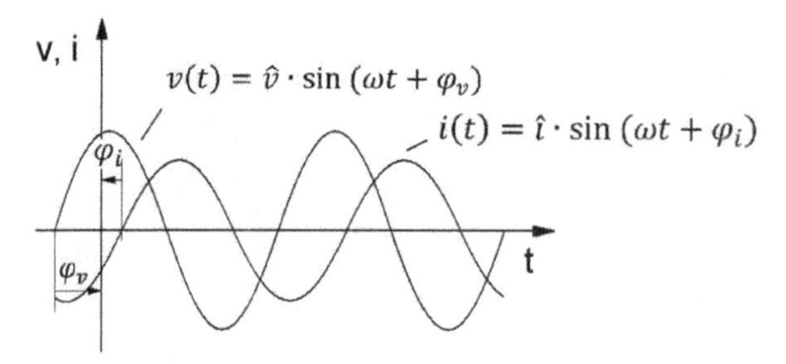

φ_v counts positive
φ_i counts negative

Complex representation

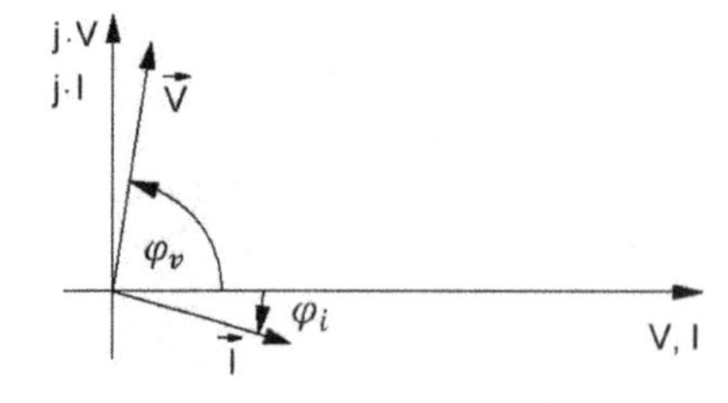

Figure 2.21: Complex representation of sinusoidal alternating quantities.

$$\underline{I} = \frac{\underline{V}}{\underline{Z}} = \frac{V}{Z} \cdot e^{j \cdot (\varphi_V - \varphi_Z)}$$

$$\underline{S} = \underline{V} \cdot \underline{I}^* \quad \underline{P} = P \quad \underline{Q_C} = -j \cdot Q_C \quad \underline{Q_L} = +j \cdot Q_L$$

2.10 Review questions

1) What is Coulomb's law and what does it constitute?
2) Describe the mechanisms of the electric current flow in metallic conductors.
3) Describe the mechanisms of the electric current flow in liquids.
4) Define the technical current direction. In which direction do the electrons flow?
5) State Kirchhoff's laws.
6) What is a mesh and what is a node?
7) Regularities regarding series connection or parallel connection of resistors?
8) What is a wye connection or a delta connection und how do you convert one into the other?
9) Sketch a voltage divider and explain the voltage divider rule.
10) What is a voltage divider used for and how high should the cross current in a loaded voltage divider be?

2.11 Exercises

EXERCISE 2.1

Three resistors with 3 Ohm, 6 Ohm and 12 Ohm were connected in series. How big is the total resistance R_s?

The same resistors were connected in parallel, how big is the total resistance R_p now?

In a mixed circuit two series resistors with 3 Ohm and 6 Ohm were connected in parallel with a 12 Ohm resistance. Calculate the total resistance and make a sketch of the equivalent circuit.

In a mixed circuit two parallel resistors with 3 Ohm and 6 Ohm were connected in series with a 12 Ohm resistance. Calculate the total resistance and make a sketch of the equivalent circuit.

EXERCISE 2.2

In a Wye connection all resistors have a resistance R of 100 Ohm, calculate the equivalent delta resistances.

In a Wye connection the resistors have following values: R_{12} = 80 Ohm, R_{23} = 250 Ohm and R_{31} = 150 Ohm, calculate the equivalent delta resistances.

In a Delta connection all resistors have a resistance R of 300 Ohm, calculate the equivalent Wye resistances.
In a Delta connection the resistors have following values: $R_{12} = 200$ Ohm, $R_{23} = 300$ Ohm and $R_{31} = 500$ Ohm, calculate the equivalent Wye resistances.

EXERCISE 2.3

When to a real voltage source, a resistor R_1 of 10 Ohm is connected a terminal voltage V_{T1} of 10V can be measured. By connecting a 6 Ohm resistor R_2 a terminal voltage V_{T2} of 9V is present. Calculate the internal resistance R_i and the constant voltage V_s.

EXERCISE 2.4

Two in series connected resistors R_1 and R_2 have a total resistance R of 400 Ohm. When a voltage of 100 V and a load resistance R of 800 Ohm is connected a voltage V_2 of 40 V can be measured at R_2.
Calculate the values of the resistors R_1 and R_2.

EXERCISE 2.5

Convert from Cartesian to exponential notation:

$$\underline{z} = 4 + j5 \qquad \underline{z} = 3 - j2 \qquad \underline{z} = -2 + j6 \qquad \underline{z} = -3 + j4$$

Convert from exponential to Cartesian notation:

$$\underline{z} = 4 \cdot e^{j20} \qquad \underline{z} = 3 \cdot e^{-j30}$$

Build the complex conjugates:

$$\underline{z} = 4 + j5 \qquad \underline{z} = 3 - j2$$

Build the sum ($\underline{z}_1+\underline{z}_2$), difference ($\underline{z}_1-\underline{z}_2$), multiplication ($\underline{z}_1 \cdot \underline{z}_2$) and division ($\underline{z}_1/\underline{z}_2$) of following complex numbers:

$$\underline{z}_1 = 3 - j2 \qquad \underline{z}_2 = -2 + j6$$

3 Fundamentals of electronics

Electronics comprise all electronic applications as well as the processes of movement of charge carriers, mostly electrons, in components of electric circuits. Electric circuits consist of electric or electromechanical individual elements (e.g. switches, displays, batteries, engines) and particular electronic components such as transistors (active components), resistors, capacitors, inductors and diodes (passive components). Since the 1960s, complete electronic circuits can be produced on one single silicon crystal. This integrated circuit (IC) has led to a constant miniaturisation and has turned semiconductor electronics the most important branch of electronics.

3.1 Semiconductor materials

!

The most important semiconductor materials are **silicon**, **germanium** and **gallium arsenide**. Their atoms are structured like a crystal lattice. Contrary to metals (=conductors), semiconductors do not have intrinsic conduction (no mobile charge carriers), but they only become conductive through energy supply (electric field, heat and light). At room temperature (293 K), however, enough electrons already move freely in a solid.

- The conductivity of semiconductor materials is highly temperature-dependent (intrinsic conductivity in silicon triples with every temperature increase of 10 K). This aspect is technically used in thermistors and is a property which is not desired otherwise.
- Semiconductor materials have to be very pure. The achieved degree of purity often is at the level of one single foreign atom in 10^{10} atoms.
- At very low temperatures, semiconductors are non-conductors.
- The specific resistance of semiconductor materials lies between the resistance of electric conductors (metals) and that of non-conductors.

Atomic structure

Semiconductor atoms form a crystal lattice. The nucleus of a silicon atom contains 14 protons and 14 neutrons. Silicon has four valence[21] electrons on the outer shell; i.e. it is quadrivalent. Each electron orbits the own and the neighbouring nucleus.

21 derived from valere = to be of value.

https://doi.org/10.1515/9783110521115-003

3.1: Examples of semiconductor materials.

Material	Application
Silicon (Si)	Diodes, transistors, integrated circuits, thyristors, solar cells
Germanium (Ge)	High frequency transistors, nuclear radiation detectors
Indium antimonide (InSb) Indium arsenide (InAs)	Magnetoresistors, hall effect sensor
Cadmium sulphide	Photoresistors, solar cells
Silicon carbide (SiC)	Thermistors, varistors, light-emitting diodes

Intrinsic conduction

At room temperature, the atoms in the crystal lattice oscillate disorderedly around their rest position (thermal movement). This leads to the breaking of a number of atomic bonds. Individual outer electrons move away from their atoms and move freely within the crystal (conduction electrons). Applying a voltage to the semiconductor crystal generates an electric field and drives the free electrons from the negative to the positive pole.

As soon as a valence electron leaves its atomic bond, a gap is created which is also called **electron hole**. These holes contribute to the current conduction because a valence electron from a neighbouring bond can fill such a hole. On the spot where the valence electron (of neighbouring bond) was before, another hole is created. The hole migrates through the whole crystal from the positive to the negative pole.

Extrinsic conduction

If a very small amount of a foreign material is added to a semiconductor material (**doping**), e.g. a single boron atom to 10^5 silicon atoms, the conductivity increases by a thousand-fold.

Foreign atoms with lower (e.g. three valence electrons) or higher valency (e.g. five valence electrons) are added to silicon. The conductivity of **extrinsic conduction** is only little **temperature dependent**.

N conductor

Free electrons (conduction electrons) are charge carriers. **Phosphorus** and **arsenic** can be used for the doping with pentavalent foreign atoms.

P conductor

Holes (electron holes) are charge carriers. We use e.g. aluminium for the doping with trivalent foreign atoms.

By adding the foreign atoms, **additional energy levels** between valence and conduction band are created. The band gap of the basic material still exists, but electrons can jump to the **newly vacated energy level** in the **p-doped** semiconductor. Hence, positive charges are created in the valence band. In the n-doped semiconductor, electrons can be elevated from the new energy level into the conduction band. In the p-doped semiconductor, current flows in the valence band; in the n-doped semiconductor, current flows in the conduction band.

Characteristics of semiconductor compared to metals

1. The conductivity highly depends on temperature and radiation effects
2. High carrier mobility and carrier velocity in the electric field
3. Conductivity through electrons (n-type conduction) and holes (p-type conduction)

3.1.1 P-n junction

At the point of contact of a p-type and an n-type conductor, a p-n junction is formed (see Figure 3.1). At the boundary from n-type to p-type conductor, electrons move from the n-type to the p-type conductor through thermal movement – no voltage is applied – and recombine with the holes there. Reversely, the holes of the p-type conductor diffuse into the n-type conductor to connect with its free charge carriers. On both sides of the boundary, the number of free charge carriers of the semiconductor crystal is reduced: At the boundary, a depletion layer is formed, which functions as an insulator.

By applying an external voltage, the p-n junction can be operated in reverse or forward direction (see Figure 39).

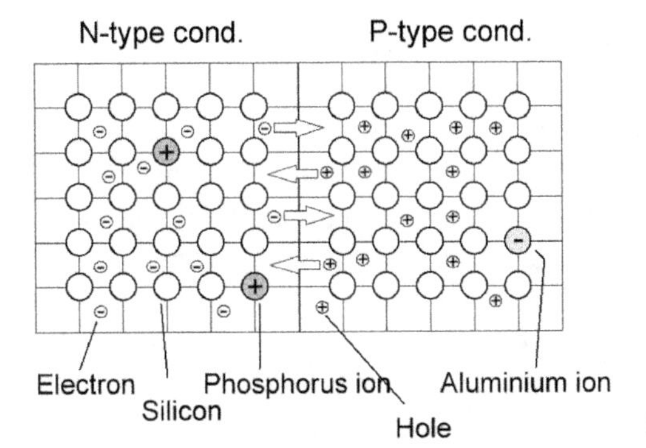

Figure 3.1: Mode of action of a p-n junction.

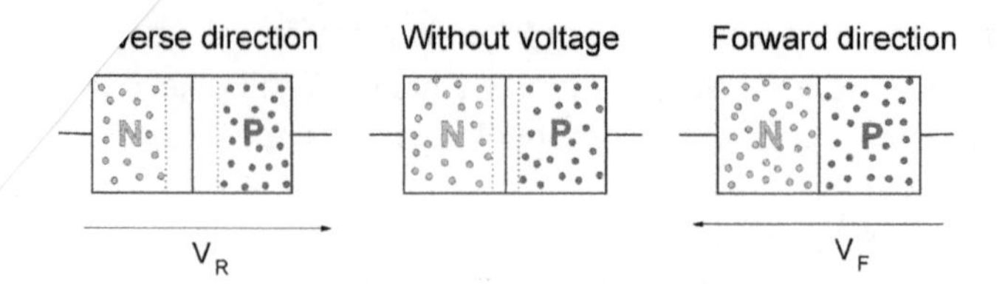

Figure 3.2: P-n junction with applied voltage.

If there is an absence of conducting electrons and holes in the boundary layer, the charges of the fixed ions exert their influence:

- The n-boundary region is positively charged; the p-boundary region is negatively charged.
- These space charge regions end further diffusion.
- The negative p-boundary layer withdraws the diffusing holes; the positive n-boundary layer withdraws the invading electrons.
- The charge in the approximately 1 μm thick boundary layer generates a diffusion voltage (0.7 V in Si).
- The p-n junction acts like a capacitor.[22]

3.2 Semiconductor diode

Diodes are resistors that have a very low resistance in forward direction, but a very high resistance in reverse direction. Therefore, they are also called valves or electrical rectifiers. The vertical line of the circuit symbol represents the cathode side of the diode,[23] the cathode side on the component itself is marked with a line around it. Diodes are used for the rectification of alternating currents. They consist of a P coating and an N coating and form a p-n junction. There are many forms of specific diodes such as Zener diodes, light-emitting diodes (LED), photodiodes, varicap diodes, laser diodes.

In a diode, a P- and an N-conductive region share a boundary and form a p-n junction. In the boundary region, a depletion layer is formed. The terminal connected to the P coating is called anode; the terminal connected to the N coating is referred to as cathode. If a positive voltage is applied to the anode (opposite the cathode), electrons are forced from the cathode into the depletion layer – the diode is conductive. In the case of a negative voltage applied to the anode (opposite the cathode), the depletion layer broadens and the diode blocks. The semiconductor

22 as charges are spatially separated and the depletion layer can be perceived as dielectric.

23 As a memory aid, we can think of the circuit symbol as an arrow pointing in the technical current direction or consider the arrow as a funnel that allows current to pass through only in this direction.

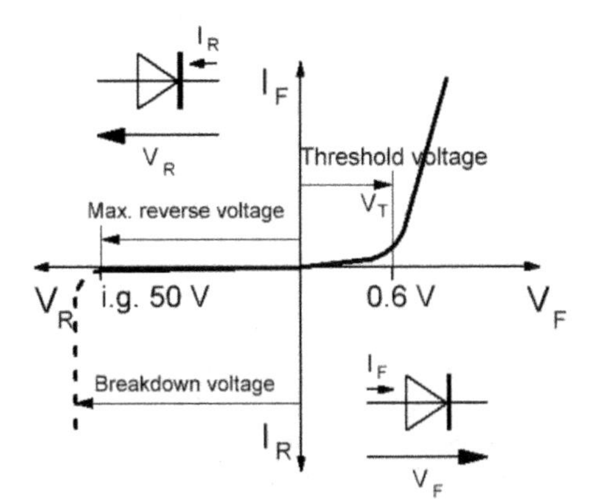

Figure 3.3: Characteristic curve and circuit symbol of a semiconductor diode.

diode conducts when polarised in forward direction. It blocks the current flow if it is polarised in reverse direction. If the diode is used in reverse direction, the P coating is at the negative pole and the N coating at the positive pole. The holes of the P coating are attracted by the negative pole; the electrons of the N coating are attracted by the positive pole. Consequently, the depletion layer, also called boundary layer, enlarges. No charge carriers can pass through the depletion layer.

Forward direction (Forward)

The positive pole of the external voltage source is connected to the anode of the diode while the negative pole is connected to the cathode. Starting at the threshold voltage V_{th} (threshold value) the diffusion voltage of the p-n junction is overcome and the forward current I_F flows.

Reverse direction (Reverse)

The negative pole of the external voltage source is connected to the anode of the diode while the positive pole is connected to the cathode. Even with increasing reverse voltage V_R only an insignificantly low reverse current I_R flows. However, above the maximum reverse voltage V_{Rmax} the reverse current increases to such a high level that it can destroy the diode. Voltages as high as or higher than the breakdown voltage destroy the diode. A temperature increase of 10 °C leads to a doubling of the reverse current.

Si power diode

Diodes are electrical valves. An ideal diode would have zero resistance in forward direction and an infinitely high resistance in reverse direction. It could be operated in forward direction with an arbitrarily high current and in reverse direction with an arbitrarily high voltage without any power loss. Diodes with a current $>1\ A$ are referred to as power diodes; they are able to rectify and shift high alternating currents. Silicon semiconductors are built into a metal case for a better heat dissipation. This case is screwed directly onto a sheet metal or onto a heat sink. Si diodes withstand maximum temperatures in the depletion layer of 150 °C. They are built for rated reverse voltages of up to 4,000 V and for forward currents of up to 1,000A.

3.2.1 Light-emitting diode (LED)

The light-emitting diode is used for display equipment, e.g. in measurement devices, watches, cars, etc. When applying external voltage in forward direction, a recombination of conducting electrons and electron holes takes place. This leads to a release of energy in form of electromagnetic radiation. The semiconductor material (e.g. GaAs) and the doping determine the emitted wavelength and therefore the light colour. A doping[24] with e.g. nitrogen (N) or phosphorus (P) results in different light colours determined by the doping level. The maximum current I_F must not exceed about 50 mA,[25] depending on the type of diode. Therefore, a series resistor R_S is always necessary for current limitation. Different light colours of diodes have different forward voltages V_F (see Figure 3.4).

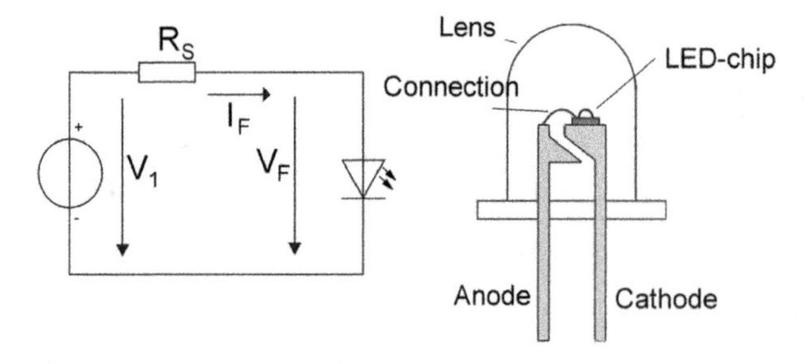

Figure 3.4: LED with series resistor.

Calculation of series resistor:

$$R_S = \frac{V_1 - V_F}{I_F}$$

R_S Series resistor
V_1 Supply voltage
V_F Forward voltage
I_F Forward current

Circuit symbol:

24 Doping = introduction of foreign atoms into the basic material.
25 ... see datasheet of the respective diode.

Table 3.2: Semiconductor materials of light-emitting diodes.

Material and doping	Colour	Wavelength in nm	Voltage V_F
GaAsSi	infrared	930	1.2 V
GaAsP	red	655	1.6 V
GaAsPN	orange	625	1.6 V
GaAsPN	yellow	590	1.8 V
GaPN	green	555	1.8 V
InGaN	blue	465	3 V
GaN/InGaN	white		3.5 V

3.3 Zener diode

The Zener diode shows a sharp bend on the reverse characteristic curve. This voltage characteristic for a Zener diode at the bend is called "Zener voltage" or breakdown voltage V_Z. A Zener diode usually operates in reverse direction.

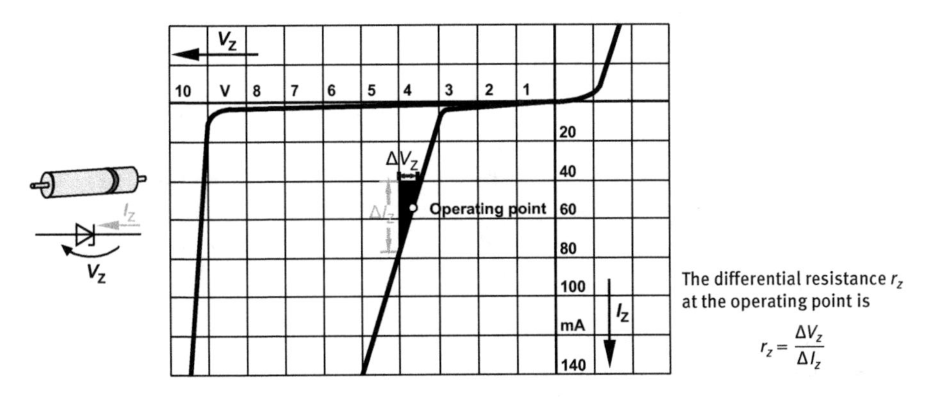

Figure 3.5: Characteristic curves of two different Zener diodes; circuit symbol.

These diodes are particularly suitable for operation in the breakdown region of the characteristic curve. In the forward region, they show a similar behaviour to a Si diode. In the reverse region, the current increases sharply after the Zener voltage V_Z has been reached. Therefore, Zener diodes in reverse direction always require a series resistor during operation (to limit the current).

Zener diodes $V_Z > 6V$ are characterised by a sharp bend, a low differential resistance r_Z (great current change I_Z causes only a small voltage change V_Z).

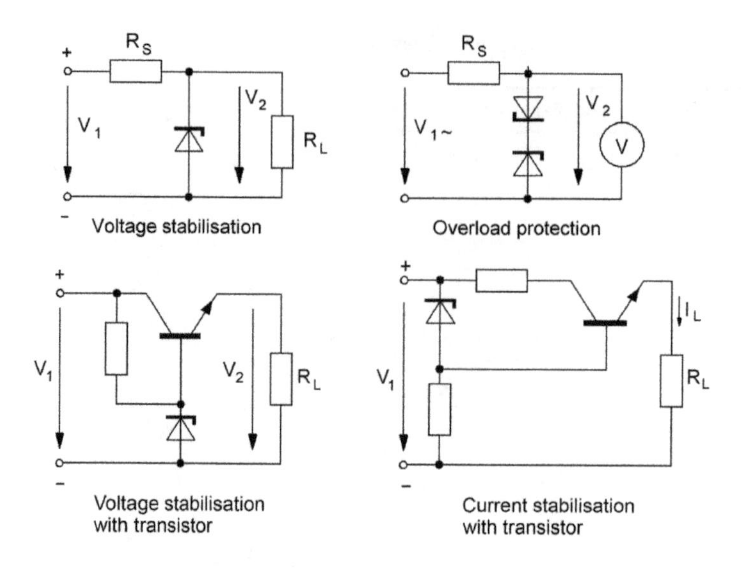

Figure 3.6: Applications of Zener diodes.

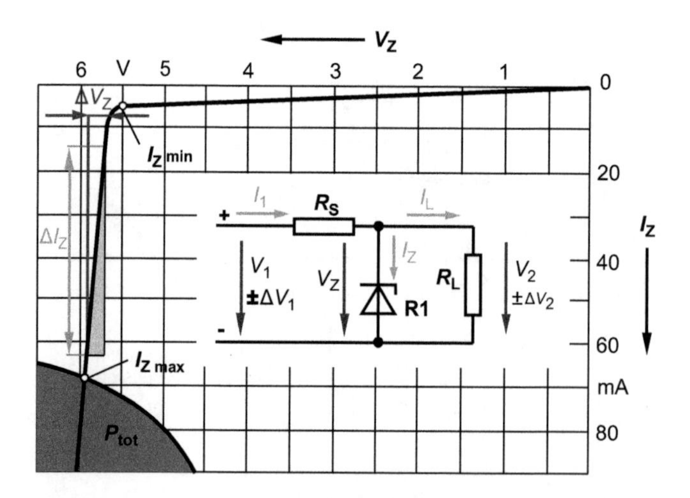

Figure 3.7: Voltage stabilisation with Zener diode (type BZX C5V6).

Mode of action of a stabilisation circuit

Let us take a look at the stabilisation circuit in Figure 3.7. A load resistor R_L is connected in parallel to a Zener diode which additionally has a series resistor R_S. The purpose of the circuit is to stabilise the voltage V_2 on the load R_L to a value as constant as possible. The load and the Zener diode are connected in parallel; therefore, the load has the same voltage as the Zener diode. The Zener voltage V_Z of the Zener diode (used in this circuit) is 5.6 V. This means that the Zener diode starts conducting from this voltage on. As long as $V_1 < V_Z$, the Zener diode blocks and the voltage $V_2 = V_1 - I_L \cdot R_S$ is applied to the load. A stabilising function is not yet given, because the voltage V_2 is linearly dependent on the voltage V_1.

In the stabilisation range of the Zener diode ($V_1 > V_Z$; $I_{Zmin} < I_Z < I_{Zmax}$) the current, however, produces a nearly constant voltage drop at the Zener diode[26] (because of the steep characteristic curve of the Zener diode). If the input voltage ($V_1 \pm \Delta V_1$) fluctuates, the voltage drop at the series resistor R_S changes accordingly[27] $\left(V_{R_S} = I_1 \cdot R_S = I_L \cdot R_S + I_Z \cdot R_S\right)$. The voltage at the load resistor R_L stays approximately the same, because the load resistor and the Zener diode are connected in parallel; the voltage applied to the load resistor is the same as the voltage drop in the Zener diode (mesh rule). The entire circuit has to be dimensioned in a way that the current through the Zener diode never exceeds I_{Zmax}, because in that case the power produced in the Zener diode would be higher than the permitted power P_{tot}. This would thermally destroy the Zener diode. The voltage V_2 can only be stabilised to the maximum value of V_1 so that I_Z is still below I_{Zmax}.

The maximum value for V_1 that we are now going to call V_{1max} is determined with the mesh rule:

$$V_{1max} = I_1 \cdot R_S + I_L \cdot R_L = I_L \cdot R_S + I_{Zmax} \cdot R_S + I_L \cdot R_L$$

3.4 Resistors

Resistors are classified according to their design. Furthermore, we distinguish between linear resistors, e.g. fixed resistors, and non-linear resistors, e.g. NTC resistors. Fixed resistors have standardised rated values (nominal values) with a permissible tolerance, which is indicated in percent of the rated value. The E series of these nominal resistors are structured in a way that they can cover every intermediate value. The designation of the E series, e.g. E12, means that there are 12 resistance values between 1.0 and 8.2 within a resistance decade (see Table 3.3).

!

26 The voltage applied to the Zener diode only changes by the value ΔV_Z.

27 The additional current flowing through the diode I_Z creates a voltage drop increased by $I_z \cdot R_V$ at R_V. At the resistor R_V, in addition to the power $I_L^2 \cdot R_V$, the power $I_Z^2 \cdot R_V$ is transformed into heat.

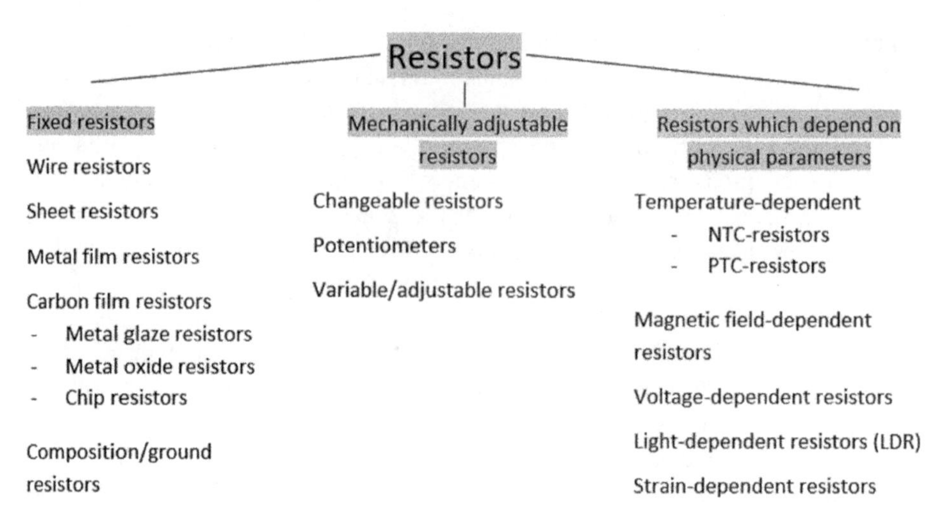

Figure 3.8: Classification of technical resistors.

The load capacity of resistors depends on how well electric heat can be emitted to the surrounding environment. High load consequently leads to large dimensions. The power ratings of resistors are given in watt at a certain temperature, e.g. 1 W at 70 °C. The resistor value of resistors is labelled with a colour code (see Table 3.4).

Table 3.3: E Series of Resistors.

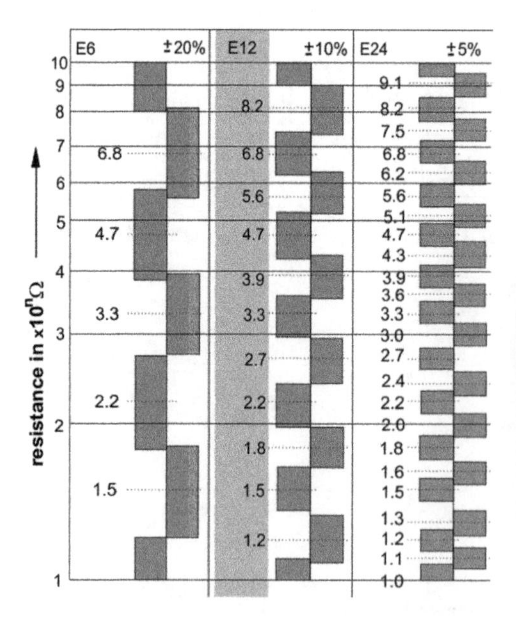

Table 3.4: Colour code of 4 ring resistors.

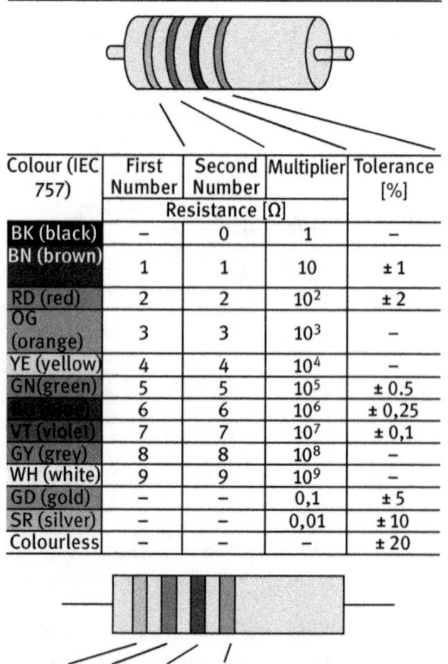

Colour (IEC 757)	First Number	Second Number	Multiplier	Tolerance [%]
	Resistance [Ω]			
BK (black)	–	0	1	–
BN (brown)	1	1	10	± 1
RD (red)	2	2	10^2	± 2
OG (orange)	3	3	10^3	–
YE (yellow)	4	4	10^4	–
GN(green)	5	5	10^5	± 0.5
[illegible]	6	6	10^6	± 0,25
VT (violet)	7	7	10^7	± 0,1
GY (grey)	8	8	10^8	–
WH (white)	9	9	10^9	–
GD (gold)	–	–	0,1	± 5
SR (silver)	–	–	0,01	± 10
Colourless	–	–	–	± 20

E.g.: 4 (YE) 7 (VT) • 10 (BN) ± 5% (GD)→ **470Ω±5%**

Wire resistors consist of a ceramic body with a resistance wire coiled around it, e.g. made from constantan. They are covered with coating, cement or glass. They are produced from 0.3Ω to 500 $k\Omega$ with power ratings up to 300W.

The resistance material of **sheet resistors** is a thin coating of crystalline coal, a noble metal or a metal oxide, e.g. on a ceramic body. Metal film resistors are temperature stable and have very low resistance tolerances (up to ± 0.005 %). Their value range is from 0.1 mΩ to 100 MΩ.

Variable resistors are produced in form of regulating resistors and turning resistors (see Figure 3.9). The three terminals are referred to as I (input), W (wiper or sliding contact) and O (output). Depending on the adjustment of the sliding contact, the tapped resistor value changes between W and O or W and I.

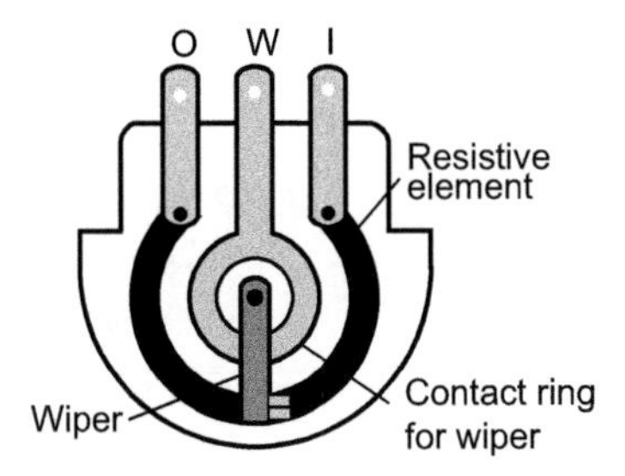

Figure 3.9: Turning resistor (potentiometer).

3.5 Semiconductor resistors

Semiconductor resistors change their resistance value depending on a physical quantity. They are often used as measuring transducers in monitoring circuits or control systems. Semiconductor resistors can be categorised into four groups:
- Voltage Dependent Resistor (VDR; varistors)
- Negative temperature coefficient resistors (NTC resistors)
- Positive temperature coefficient resistors (PTC resistors)
- Magnetic dependent resistors (magnetoresistors)

3.5.1 Voltage dependent resistor (varistors)

The resistance of a varistor (**Var**iable Res**istor**, **V**oltage **D**ependent **R**esistor VDR) is high at low applied voltage and low at high voltage.
- Non-linear resistance
- Resistance R decreases with voltage V

Production process: Silicon carbide powder with binding agents pressed into the form of sticks or discs and sintered. Charges on grain boundaries impede the current flow and cause "contact resistance".

The contact resistance of the metal oxide grains (ZnO) or silicon carbide grains (SiC) is voltage-dependent. The current through the VDR grows with increasing voltage, slowly at first and then ever more rapidly. The characteristic curve is non-linear but symmetrical to the origin. ZnO varistors almost have switching characteristics and can limit voltages. During an overvoltage pulse the varistor abruptly reduces its resistor value from several megaohm to just a few ohm.

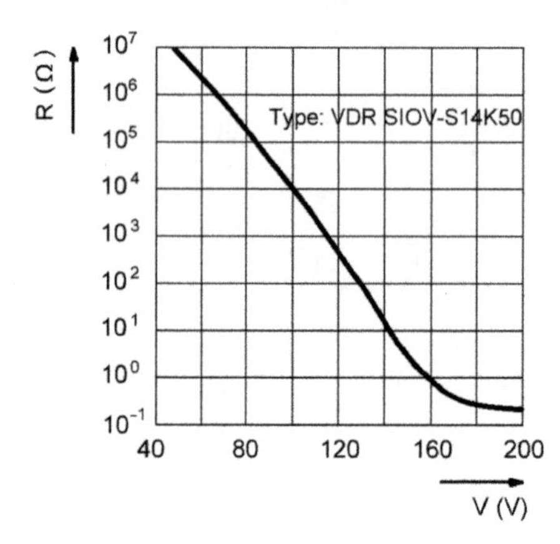

Figure 3.10: Resistance characteristic curve of a varistor.

Area of application of VDRs:
- They are often used to suppress overvoltage pulses, rarely for voltage stabilisation.
- Protection of components sensitive to overvoltage like diodes, transistors, thyristors or integrated circuits
- Protection of switches against combustion
- Protection from overpotential peaks that originate from supply lines

3.5.2 Negative temperature coefficient resistors (NTC resistors)

A thermistor (NTC[28]) is a temperature dependent resistor with a resistor value that decreases with increasing temperature (hence the name). It has a negative temperature coefficient. NTCs are made of metal oxides (mainly MgO and TiO_2). After

28 **N**egative **t**emperature **c**oefficient.

grinding and mixing, the mass is pressed into the desired shape in a steel mould and is then sintered at temperatures up to 1,600 °C.

A change in temperature can occur **externally** through the ambient temperature (externally heated thermistors) or **internally** through the heating by means of current flowing through (self-heating thermistor).

Externally heated thermistors are operated in the high-resistance range of the characteristic curve where the current flowing through the component is so low that the thermistor is hardly heated at all. With fixed resistors that are connected in parallel or series to the thermistor, one can linearise/straighten the resulting characteristic curve.

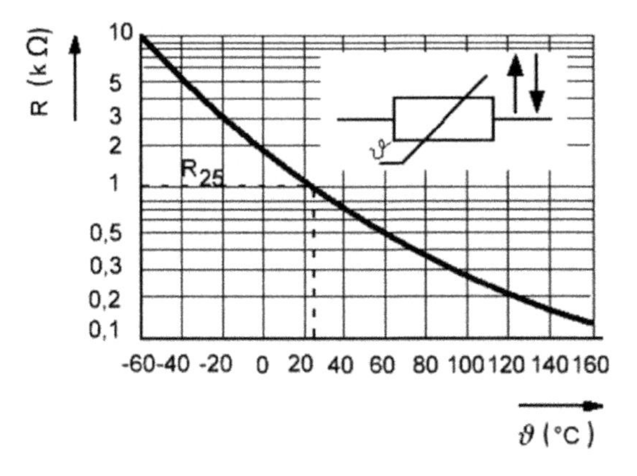

Figure 3.11: Resistance characteristic curve and circuit symbol of a thermistor.

Properties

- Area of application: Suitable for temperatures up to approximately 150 °C.
- Due to their small design the NTCs rapidly react to temperature changes.
- The minimal differences in temperature can be determined.
- Disadvantages:
  - Non-linear characteristic curve
  - Characteristics specific to material and production rule out a standardisation of NTCs. Therefore, products of different manufacturers are interchangeable only to a limited extent.
- Advantage: cost-effective

Areas of application:

Applications of externally heated thermistors:
- Temperature measurement in devices and systems (e.g. air conditioning system)
- Resistance thermometer

- Power measurement of microwaves
- Temperature compensation of components

Applications of self-heating thermistors:

Inrush current limiting, Christmas tree lighting, relay with pick-up delay

3.5.3 Positive temperature coefficient resistors (PTC resistors)

PTC resistors[29] are temperature dependent resistors with a positive temperature coefficient. All metals have a positive temperature coefficient; hence, they also can be referred to as PTC resistors. We, however, are referring to PTC resistors from semiconductor materials in this section.

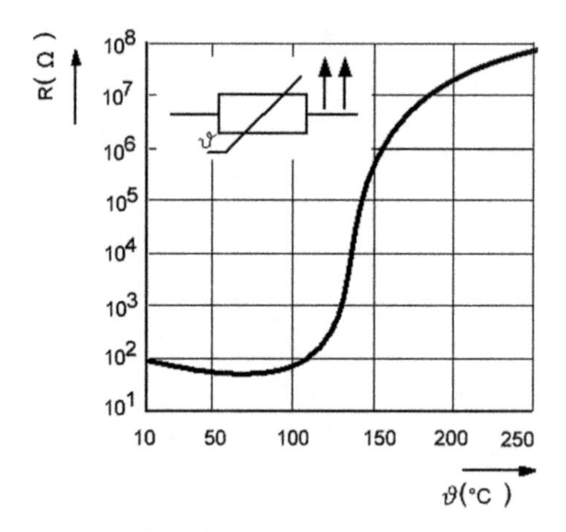

Figure 3.12: Characteristic curve and circuit symbol of a PTC resistor.

PTC resistors from semiconductor materials show the following behaviour:

Initially, the resistance decreases with increasing temperature until the outlet temperature at the smallest resistance R_{min}, as it happens with every semiconductor. Heated further, the resistance coefficient increases sharply by a thousand-fold until the terminal resistance is reached. In a limited temperature range, the resistance of a PTC resistor increases sharply.

Externally heated PTC resistors may only be heated slightly by the current passing through

29 PTC = **p**ositive **t**emperature **c**oefficient.

Applications for externally heated PTC resistors:

- Industrially, PTC resistors are used as **temperature sensors** in form of platinum measuring resistors (e.g. Pt100 and Pt1000) up to temperatures of approximately 200 °C (however, there are types used up to 450 °C). Advantages over NTC resistor: approximately linear characteristic curve, standardisation (measuring resistors of the same type are interchangeable)

Applications for self-heating PTC resistors:

- Overload protection or protection against short circuits of small engines and relay coils with PTC resistors connected in series.
- Flow measurement in liquids
- Fluid level sensor
- Limit monitoring indicator of overfilling protection in fuel-oil tanks
- Thermally stable heating element (if the temperature falls, more current and power are carried due to a lower resistance, which means that the heating increases)

3.5.4 Magnetic dependent resistors (MDR[30])

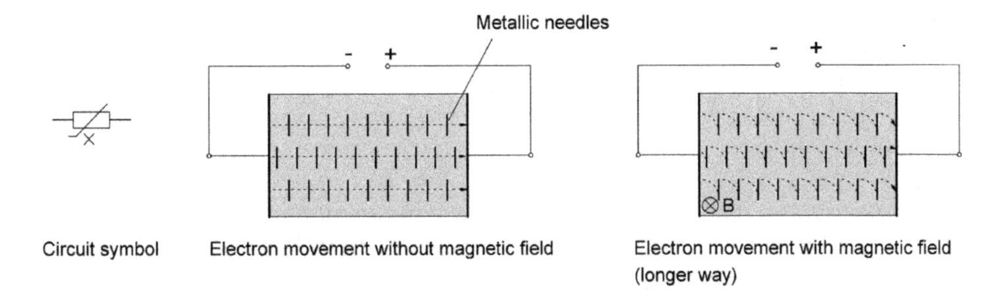

Figure 3.13: Circuit symbols and operating principle of a magnetic dependent resistor.

Magnetic dependent resistors are semiconductor resistors with a resistor value that is influenced by a magnetic field. A current-carrying conductor changes its resistance in a sufficiently strong magnetic field. The cause for this is the deflection of mobile charge carriers through the magnetic field force (Lorentz force) and the hence occurring extension of the current path (see Figure 3.13). The resistance control through the impact of the magnetic field practically happens without inertia. The magnetoresistor is able to follow rapid field changes with frequencies of up to several MHz.

30 **m**agnetic field **d**epending **r**esistor.

Application
- Measuring magnetic fields
- Measuring direct currents using current clamps
- Contactless variable resistors
- Contactless switches
- Rotational speed measurement

Example: The magnetoresistor MR 30D 250E has a basic resistance $R_0 = 250\Omega$ at $B = 0T$. If the magnetic flux density is increased to 1 T, the resistance grows by the factor of 15.

3.6 Hall effect sensor

The Hall effect sensor creates voltage by means of a current and a magnetic field (Hall voltage). The Hall effect sensor uses the Hall effect.[31] If the current is known, we can measure the magnetic field, as the magnetic field is proportional to the generated voltage. The Hall effect sensor (see Figure 3.14) is a thin semiconductor plate with four terminals. Due to an applied voltage, a current I_1 flows through the semiconductor plate. The plate is additionally affected by a magnetic field with the induction B; the voltage V_H is generated as a result of the force effect on the charge carriers (Lorentz force) at both measuring terminals. The Hall voltage increases with the intensity of the control current I_1 and the magnetic flux density B. An open circuit Hall effect sensor $I_2 = 0$ has the open circuit Hall voltage V_{20} $(V_{20} \sim B \cdot I)$. When $B = 0$, $V_H = 0$ (zero point error).

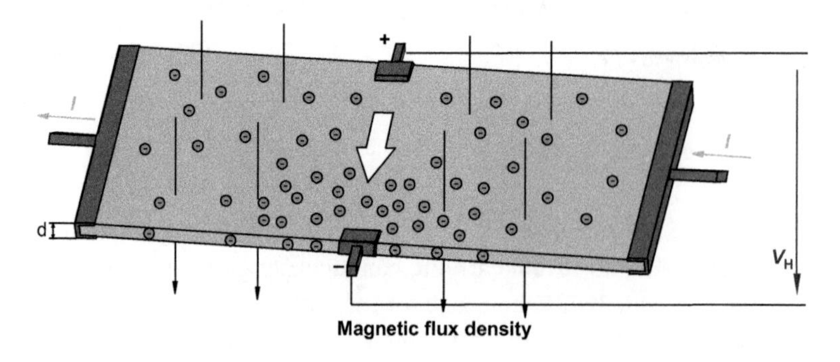

Figure 3.14: Circuit symbols and function of a Hall effect sensor.

31 The Hall effect (named after Edwin Hall; US physicist; 1855 – 1938) occurs in a current-carrying conductor that is located in a magnetic field. Thereby an electric field builds up which is perpendicular to the current direction and the magnetic field and compensates the Lorentz force affecting the electrons.

Application:
- Measuring magnetic fields
- Potential-free current measurement
- Contactless signal generators
- E.g. seatbelt buckles, door locking system, pedal position sensing
- Control of ignition timing as pulse generator

3.7 Transistor[32]

Transistors are active semiconductor components that are used for voltage adjustments, for switching and controlling as well as for the amplification of voltages and currents, without performing mechanical movements.

Table 3.5: Main areas of application of transistors.

Voltage adjustments	Switching	Amplifying
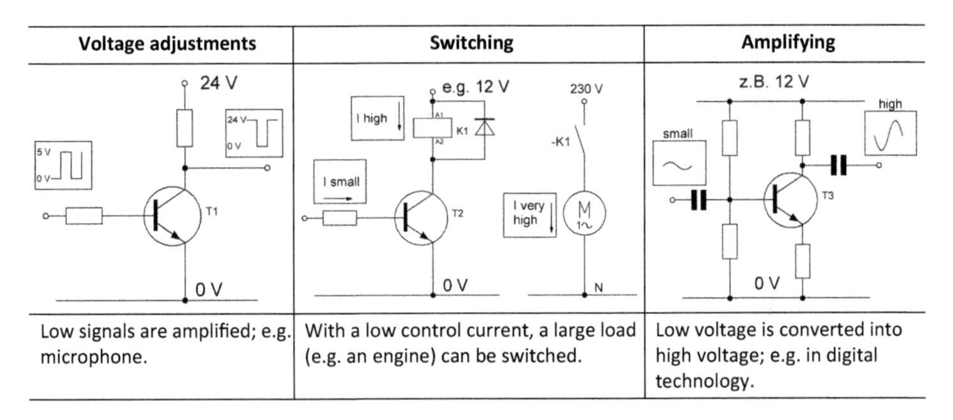		
Low signals are amplified; e.g. microphone.	With a low control current, a large load (e.g. an engine) can be switched.	Low voltage is converted into high voltage; e.g. in digital technology.

There are two main groups of transistors: **bipolar transistors** (often produced using silicon) and unipolar transistors (**field-effect transistor, FET**) that differ in the way of control.[33] When speaking of "transistors", we usually mean bipolar transistors. They are referred to as "bipolar" because electrons and holes are involved in the charge transport.

The current flows through alternating N- and P-conductive regions: emitter E, base B and collector C. The two p-n junctions can be represented[34] as two opposing diodes

32 The term "transistor" is a blend of the words transfer und resistor. A transistor can also be seen as an electrical resistor controllable through electric current.

33 Bipolar transistors are controlled through a current, while unipolar transistors are controlled through an electric field – caused by voltage.

34 In reality, a transistor cannot be built from two diodes.

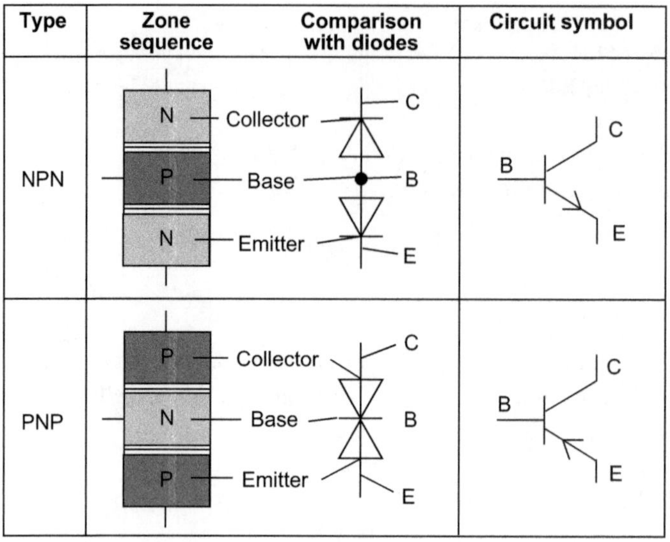

Type	Zone sequence	Comparison with diodes	Circuit symbol
NPN	N – Collector P – Base N – Emitter	C B E	B, C, E
PNP	P – Collector N – Base P – Emitter	C B E	B, C, E

Figure 3.15: Structure and circuit symbols of bipolar transistors.

(see Figure 3.15). Transistors consist of three stacked semiconductor layers. Depending on the zone sequence, the transistor is either a PNP or a NPN type (see Figure 3.15). As

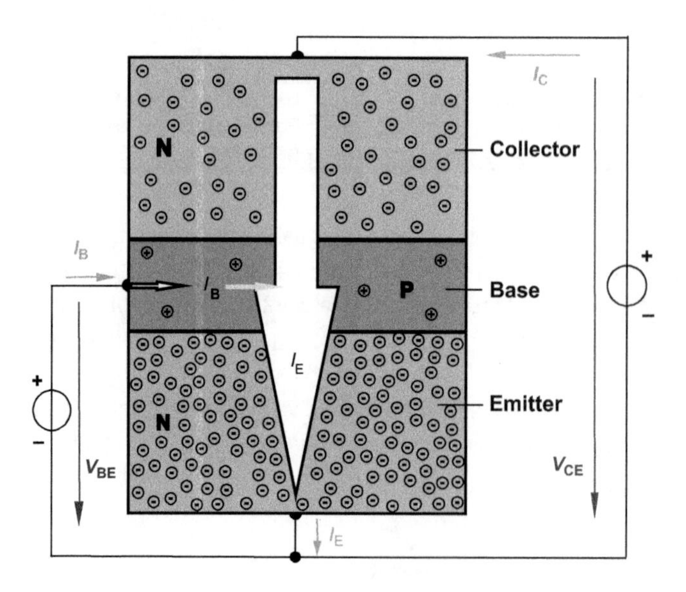

Figure 3.16: Functional schematic of a NPN transistor.

single transistors, we mainly use NPN transistors. The three semiconductor regions are contacted, the terminals lead to the outside. The region in the middle is called base, the two outer layers are referred to as emitter and collector. The emitter sends out charge carriers, which are collected by the collector. The base-emitter path is polarised in forward direction during transistor operation; the base-collector path is polarised in reverse direction. The arrow of the emitter indicates the technical current direction.

A low base current I_B controls a high collector current I_C in a (bipolar) transistor. This is referred to as current amplification. For controlling, only a low electric power is needed.

The output current I_C is proportional to the base current I_B over a long period and hence accordingly controllable. For this freely determined operating point, we consider the following quantities:

Direct current amplification B	**Alternating current amplification β**
$B = \frac{I_C}{I_B}$	$\beta = \frac{\Delta I_C}{\Delta I_B}$
I_C Collector current in A	ΔI_C Collector current change in A
I_B Base current in A	ΔI_B Base current change in A

Direct current input resistance R_{BE}	**Direct current output resistance R_{CE}**
$R_{BE} = \frac{V_{BE}}{I_B}$	$R_{CE} = \frac{V_{CE}}{I_C}$
V_{BE} Base-emitter voltage in V	V_{CE} Collector-emitter voltage in V
I_B Base current in A	I_C Collector current in A

The emitter region is highly doped, the collector region a little less so. The exceptionally thin base layer (a few µm) only contains a very low number of foreign atoms. In case of a flowing base current, charge carriers flood the thin base layer coming from the emitter (e.g. electrons in an NPN transistor). Since the base layer is lowly doped, only a small amount of electrons can recombine with the holes. The majority of charge carriers are driven into the collector by the electric field of the base-collector depletion layer. This is how the high collector current (10 to 500 times higher) is generated = direct current amplification.

3.7.1 Basic amplifier circuits

Amplifiers are used to faithfully amplify signals without distortion. In accordance with the three terminals of the transistor, we differentiate between emitter, collector, and base circuits (see Table 3.6). To keep the load on the source, which is to be amplified, low, the input resistance R_{IN} should be as high as possible. A low output resistance R_{OUT} allows the highest possible output current.

$$V_v = \frac{V_{2\sim}}{V_{1\sim}} \qquad V_i = \frac{I_{2\sim}}{I_{1\sim}} \qquad V_p = \frac{P_{2\sim}}{P_{1\sim}} = V_v \cdot V_i$$

Voltage amplification factor **Current amplification factor** **Power amplification factor**

Table 3.6: Basic amplifier circuits of bipolar transistors.

	Emitter circuit	Collector circuit	Base circuit
Circuit	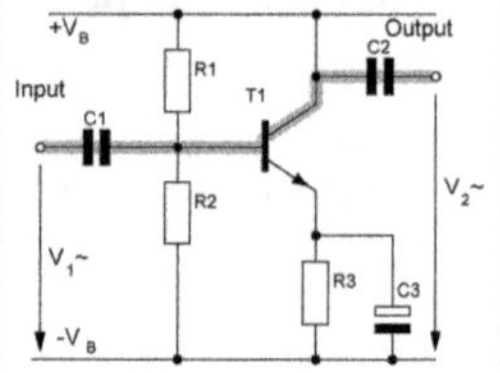	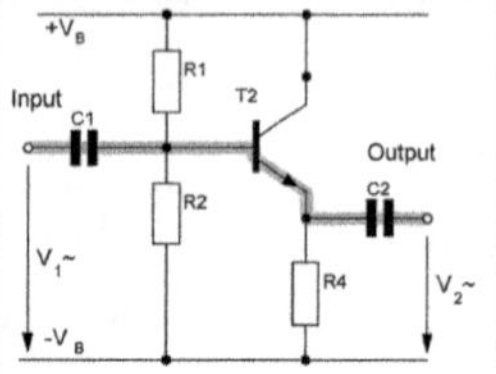	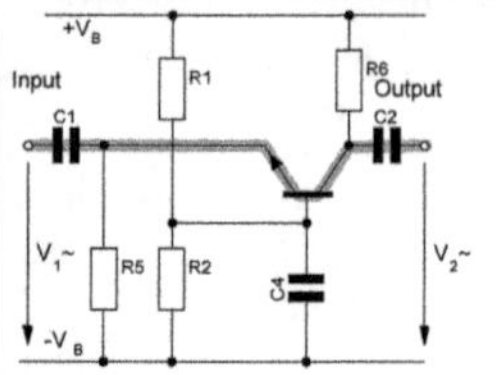
V_v	high, e.g. 200	low (<1)	high, e.g. 200
V_i	high, e.g. 200	high, e.g. 200	low (<1)
V_p	very high, e.g. 4,000	high, e.g. 200	high, e.g. 200
R_{IN}	medium, e.g. 5 kΩ	high, e.g. 50 kΩ	low, e.g. 50 Ω
R_{OUT}	high, e.g. 20 kΩ	low, e.g. 100 Ω	high, e.g. 50 kΩ
φ	180°	0°	0°
Application	LF amplifier	LF input amplifier	HF amplifier

To understand how the amplification process happens in the transistor, we take a look at the four quadrant characteristic field of a transistor (see Figure 3.17). The input voltage v_1 (on the bottom right) causes a change of the base current I_B. An altered base current leads to change in the collector current I_C, which in turn determines the output voltage v_2 through the operating load line.

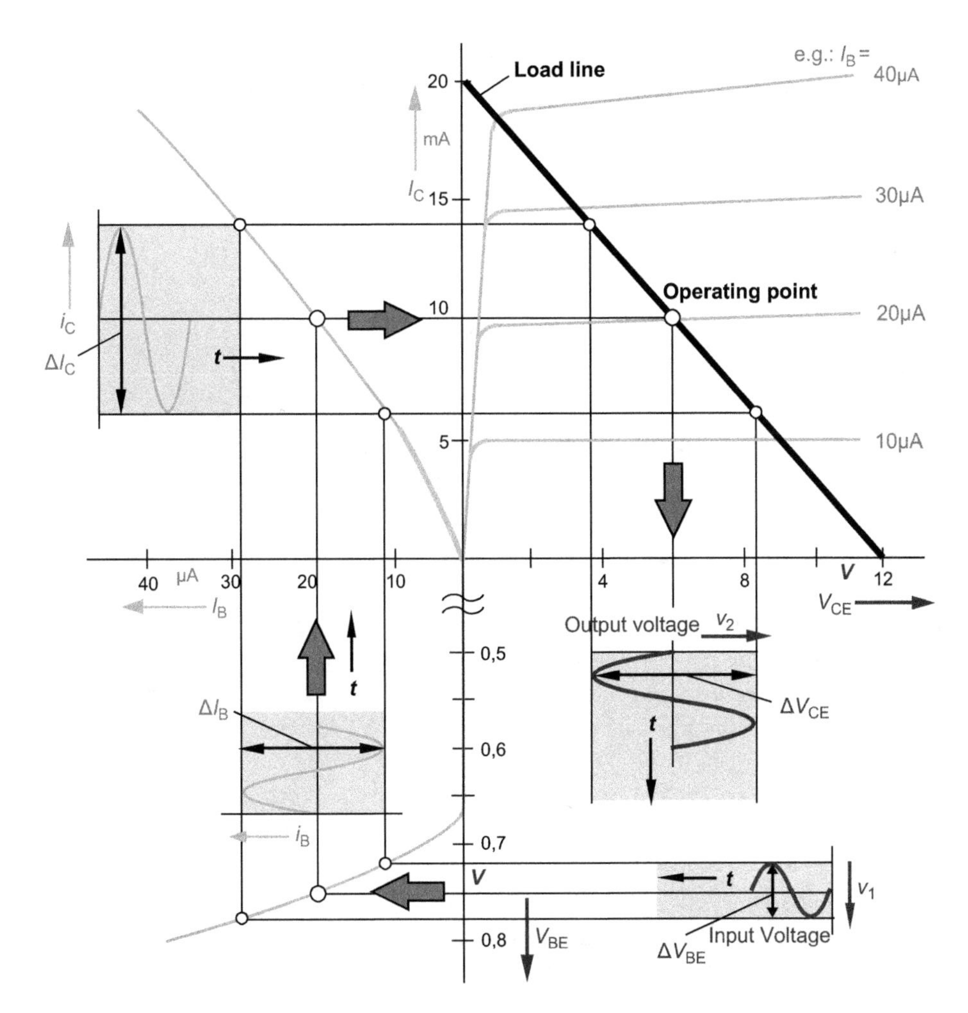

Figure 3.17: Amplification process of an alternating voltage in the four quadrant characteristic field.[35]

3.8 Optocouplers

Optocouplers allow the transfer of signals between two **galvanically isolated**[36] circuits. The two circuits are "optically coupled" (hence the name). The load circuit does not influence the control circuit and vice versa. Within the optocoupler, the

35 The four quadrant characteristic field of a transistor represents the connection of the characteristic quantities of said transistor.

36 Galvanic isolation (also referred to as galvanic decoupling) means the absence of an electrically conductive connection between two circuits. However, electric power or electrical signals can still

light-emitting diode is used as the sender and e.g. a phototransistor acts as receiver. A light-proof casing shields the sender and receiver from external influences.

Application:
- In automation technology
- In electronic load relays
- To separate potentials in PLC inputs

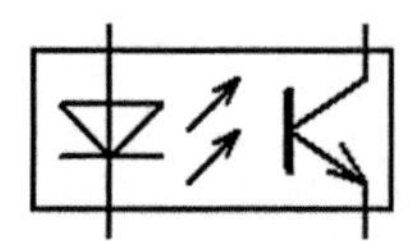

Figure 3.18: Circuit symbol of optocouplers.

3.9 Operational amplifier

Operational amplifiers (also referred to as "op-amp"[37] or "opamp") are produced as ICs in integrated technology and are considered one of the most important components in analogue electronics. Only a few external components are required to construct a variety of circuits that electrically realise mathematical operations (hence the name "operational amplifier"). In this way, it is possible to build adders, subtractors, multipliers, integrators, dividers as well as logarithmic amplifiers with only one (or several) op-amp(s).

Op-amps are used, for example, as basic component in amplifiers, instrumentation amplifiers, impedance converters, comparators, Schmitt triggers, as multivibrator, controller, for flip-flops, digital-to-analogue converters, constant current and constant voltage generators, active filters and for analogue signal generators. Operational amplifiers have a positive (non-inverting, "+") and a negative (inverting, "–") input as well as an output. They amplify the voltage difference V_D between the two inputs.

be transferred through a magnetic field or light. In galvanic isolation, also electric potentials are separated from each other; circuits are potential-free with regard to one another.

37 Operational amplifier.

Table 3.7: Characteristic values of an op-amp for different types.

		Ideal
High amplification	$V_0 = 10^3 \ldots 10^6$	∞
Great input impedance	$Z_E = 10^5 \ldots 10^{14}\Omega$	∞
Small output impedance	$Z_A = 10 \ldots 200\Omega$	0
Output current	I_{Amax} : up to 100 mA	∞
Slew rate[38]	$SR = 0,5\frac{V}{\mu s} \ldots 600\frac{V}{\mu s}$	∞

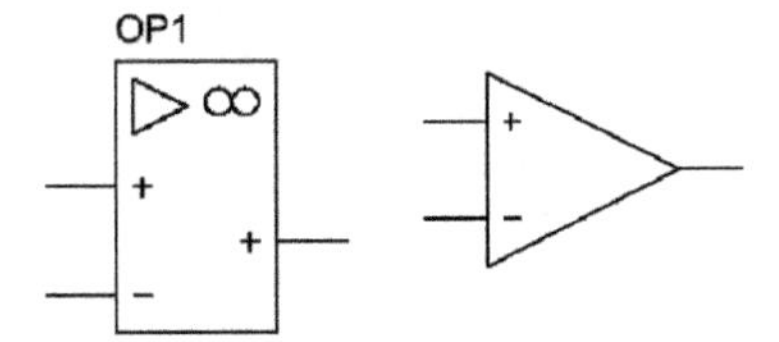

Figure 3.19: Circuit symbol of the operational amplifier.

Golden rules:
When calculating circuits with operational amplifiers, the following rules may be applied based on the assumption that an ideal operational amplifier is used:
a) The output is positive if the voltage applied to the non-inverting input is higher than the voltage at the inverting input.
b) Provided a feedback is built into the circuit, the voltage at the output will change until the voltage between the inverting and the non-inverting input 0 V is ($V_D = 0$ V).
c) There is no current flow, neither in the inverting nor in the non-inverting input.

3.9.1 Basic circuits of an op-amp

The following sections introduce three important basic circuits:

Inverting amplifiers
In an inverting operational amplifier, part of the output voltage is inverted via the feedback resistor R_F and led back to the inverting input. Due to the positive input voltage V_i, a differential input voltage V_D occurs. V_D is amplified through the voltage

38 Slew rate.

amplification V_O. The negative output voltage V_o increases[39] and reduces V_D through the resistors R_F. V_o increases until V_D practically equals zero and the op-amp is adjusted.

The differential input voltage V_D is now zero, hence, the input "–" with the node S (virtual ground point) has the same ground potential as the non-inverting input "+". Therefore, the following applies: $I_F + I_e = 0$. Through $I_F = V_o/R_F$ and $I_e = V_i/R_i$ we calculate the amplification factor V considering the ratio of resistances R_F/R_i.

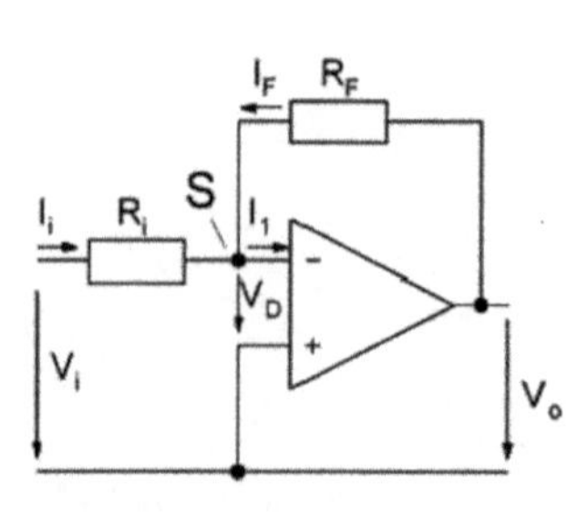

$$V_a = -\frac{R_F}{R_i} \cdot V$$

$$V = \frac{V_o}{V_i} = -\frac{R_F}{R_i}$$

V_i	Input voltage
V_o	Output voltage
R_F	Feedback resistor
R_i	Input resistance
V	Voltage amplification factor

Figure 3.20: Inverting operational amplifier.

Non-inverting amplifier

The non-inverting amplifier has a high input resistance (e.g. 10 $M\Omega$) and a significantly lower output resistance (e.g. 0.1 Ω). The input voltage V_i and the output voltage V_o share the same algebraic sign. The non-inverting amplifier is therefore well-suited as **measuring amplifier**. In this circuit (see Figure 3.21) – with the approximation $V_D \approx 0$ and $I_1 = 0$ – the voltage divider connected to the total voltage V_o and the partial voltage V_i resulting from the resistors R_F and R_S is almost completely unloaded. The following applies:

$$I_1 = 0 \text{ leads to } I_F = I_C$$

$$\text{for } I_F = \frac{V_o}{R_C + R_F} \text{ and } I_C = \frac{V_i}{R_C}$$

by equating, the following is obtained:

$$\frac{V_o}{R_C + R_F} = \frac{V_i}{R_C} \Rightarrow V_o = V_i \cdot \left(1 + \frac{R_F}{R_S}\right)$$

39 . . . with the slew rate . . . .

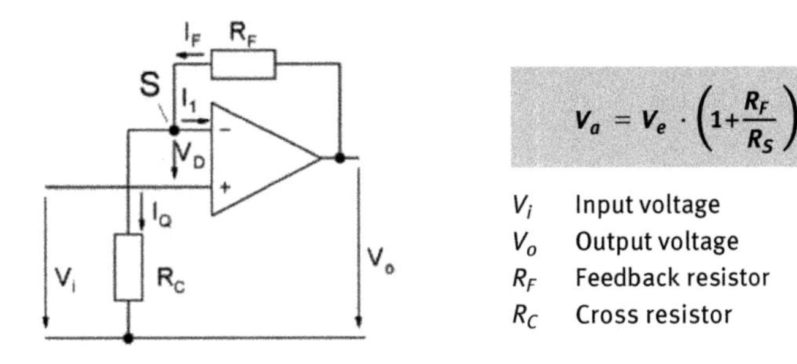

$$V_a = V_e \cdot \left(1 + \frac{R_F}{R_S}\right)$$

V_i	Input voltage
V_o	Output voltage
R_F	Feedback resistor
R_C	Cross resistor

Figure 3.21: Non-inverting amplifier.

Impedance converter (voltage follower)

In an impedance converter, the output voltage V_o equals the input voltage V_i. The input resistance is very high; the output resistance is very low ($R_F = 0$ and $R_C = \infty$). The impedance converter adjusts the high internal resistance of a signal source to a low load resistance.

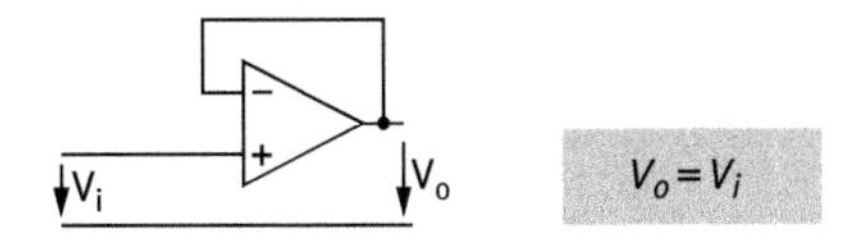

$$V_o = V_i$$

Differential amplifier

Operational amplifiers that are connected as differential amplifiers are of great significance in measurement technology. They are used e.g. in electronic temperature measurement to amplify the differential voltage of a bridge circuit. Often a differential amplifier with $R_1 = R_i$ and $R_2 = R_F$ is selected.

With the approximation $I_1 \approx 0$ and $V_D \approx 0$ we obtain:

1. **For $V_{i1} = 0$:**

$$V_o = V_{R2} \cdot \frac{R_i + R_F}{R_i} \quad \text{and} \quad V_{R2} = V_{i2} \cdot \frac{R_2}{R_1 + R_2}$$

$$V_o = V_{i2} \cdot \frac{R_2}{R_1 + R_2} \cdot \frac{R_i + R_F}{R_i}$$

2. **For $V_{i2} = 0$:**

$$V_o = -V_{i1} \cdot \frac{R_F}{R_i}$$

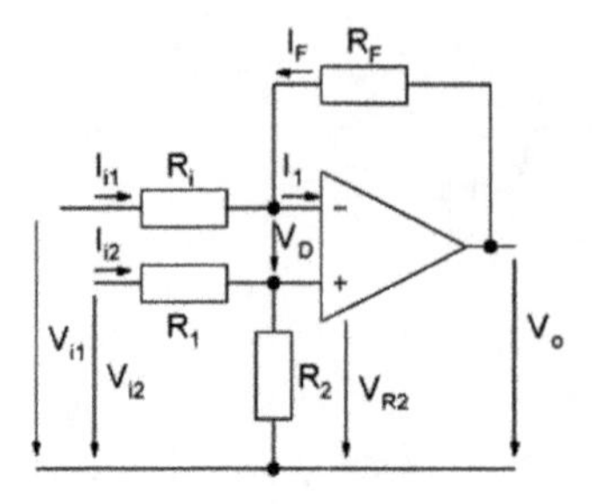

$$V_o = (V_{i2} - V_{i1}) \cdot \frac{R_F}{R_i}$$

Figure 3.23: Differential amplifier.

Overlap: with $R_1 = R_i$ and $R_2 = R_F$:

$$V_o = V_{i2} \cdot \frac{R_2}{R_1 + R_2} \cdot \frac{R_i + R_F}{R_i} - V_{i1} \cdot \frac{R_F}{R_i}$$

$$\Rightarrow \mathbf{V_o = -(V_{i1} - V_{i2}) \cdot \frac{R_F}{R_i}}$$

Integrators / Differentiators

In an integrator (see Figure 3.24) the capacitor C_F is located in the negative feedback branch. The integrated circuit is used e.g. to generate sawtooth voltages and in digital-to-analog conversion.

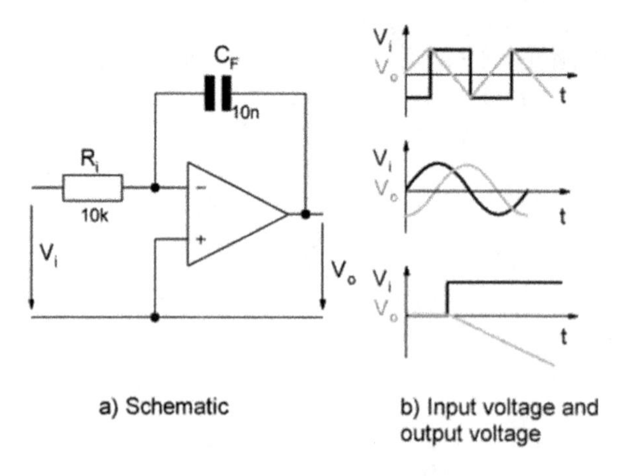

Figure 3.24: Operational amplifier used as integrator.

The differentiating circuit (see Figure 3.25) is equipped with a capacitor C_i at the input. Current can only flow through C_i if the input voltage V_i changes.

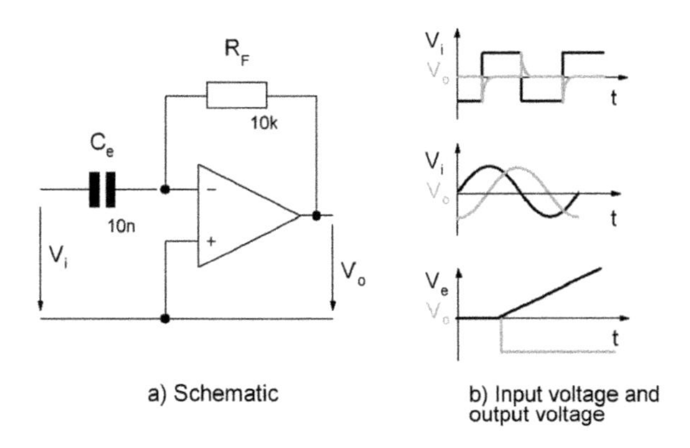

Figure 3.25: Operational amplifier used as differentiator.

3.10 Thyristor

A thyristor is a component with bistable[40] behaviour. It consists of four serially connected semiconductor layers that are differently doped in order to create three depletion layers. The two outer layers function as terminals connected to the anode and cathode, an inner layer is connected to the control terminal (gate). A thyristor[41] (or SCR[42]) consists of four consecutive semiconductor regions (PNPN) and is either formal or real (see Figure 3.26 b) due to the combination of an NPN- and a PNP-type transistor. A positive gate potential opposite the cathode overrides the blocking state if the gate current is high enough.

When connected to a gate voltage, the thyristor "ignites" and behaves like a diode in forward direction. If a brief voltage pulse is applied to the gate of the thyristor, it acts like a diode. Only after the current flowing between the anode and cathode drops below the holding current I_H the thyristor returns to blocking state.

Rated voltages 50–10,000 V; rated currents 0.4–6,000 A

- P-gate thyristors are commonly used in practice; the outer P layer functions as anode, the outer N layer is the cathode and the inner P layer is the gate. Gate current I_G floods the inner P conductor, which leads to the reduction of the depletion layer.
- The thyristor acts like a diode as soon as a gate current flows. The forward voltage V_F is 0.6–3 V; the gate voltage required for ignition V_{GI} is 0.6–2.5 V.

---

40 Bistable: There are two possible stable states; changing from one state into the other is only possible due to an external impulse (ignition pulse in a thyristor).

41 The term "thyristor" is derived from the terms thyratron and transistor. A thyratron is a gas-filled tube or discharge tube that has similar properties to a thyristor.

42 Silicon Controlled Rectifier.

- A thyristor has three depletion layers. If a voltage is in effect between the anode and the cathode, at least one of the three layers is polarised in reverse direction.
- The conducting direction is the direction of voltage, which results in only one p-n junction being polarised in reverse direction in the thyristor.
- The direction where two depletion layers are polarised in non-conducting direction is called reverse direction.

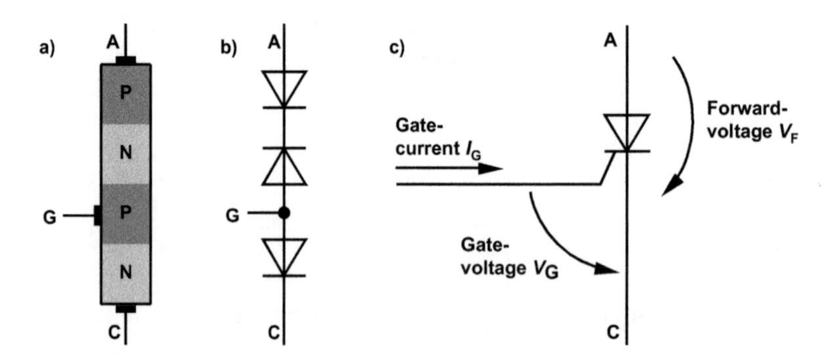

Figure 3.26: Thyristor a) structure, b) structure with 2 transistors, c) circuit symbol with characteristic values.

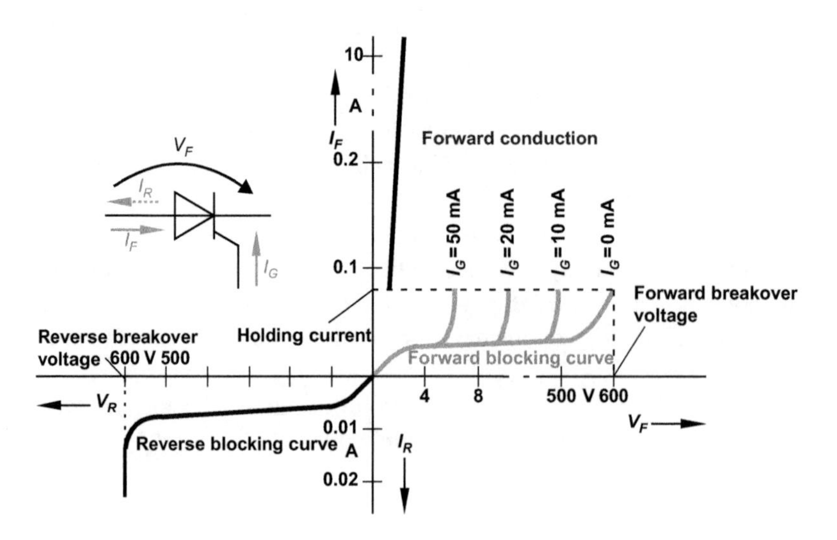

Figure 3.27: Characteristic curve of a thyristor.

Application: Phase-fired control (e.g. in dimmer switches, speed control of universal motors, etc.), rectifiers, contactless switches, H bridges

3.11 Rectifier circuit

Rectifiers are used in electrical engineering and electronics to convert alternating voltage into direct voltage.

Two-pulse bridge circuit (B2U)

The two-pulse bridge circuit is the most commonly used rectifier circuit.

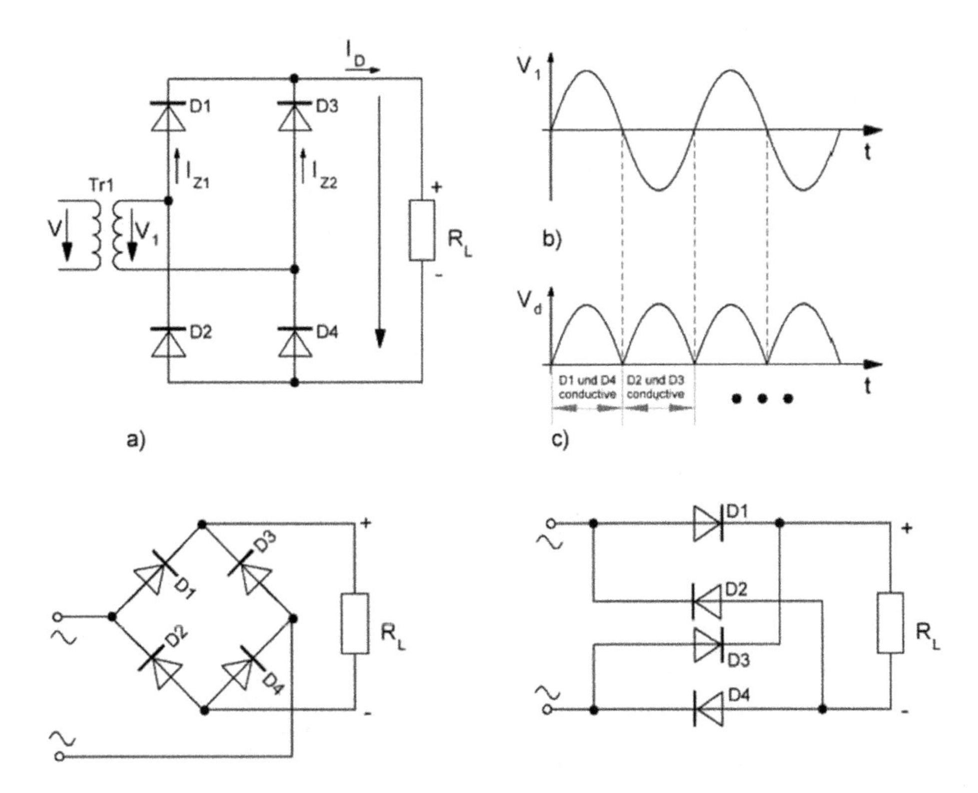

Figure 3.28: a) Two-pulse bridge circuit (B2U); b) Input voltage; c) Output voltage; d) Further representations.

Function: If the voltage V_1 is positive (see Figure 3.28b), the branch current I_{B1} flows through the diode D1, the load resistor R_L and the diode D4. With negative voltage V_1 the branch current I_{B2} passes through the diode D3, the load resistor R_L and the diode D2. The load receives a pulsating direct voltage V_d. The two-pulse bridge circuit uses both half cycles of alternating voltage. Therefore, the arithmetic mean of the ideal open circuit direct voltage V_{di} is twice as high as the arithmetic mean in the one-pulse one-way circuit E1U. It amounts to 90% of the effective value of the alternating voltage V_1.

$$V_{di} = 0.9 \cdot V_1 \qquad P_T = 1.23 \cdot P_d$$

V_{di} Open circuit direct voltage in V
P_T Transformer power in W
P_d Direct-current power in W

3.11.1 Rectified value (arithmetic mean)

The rectified value represents the arithmetic mean of an alternating voltage or alternating current. It is the mean over a period. The rectified value of an alternating quantity is always zero; the rectified value of a pulsed quantity can be positive or negative.

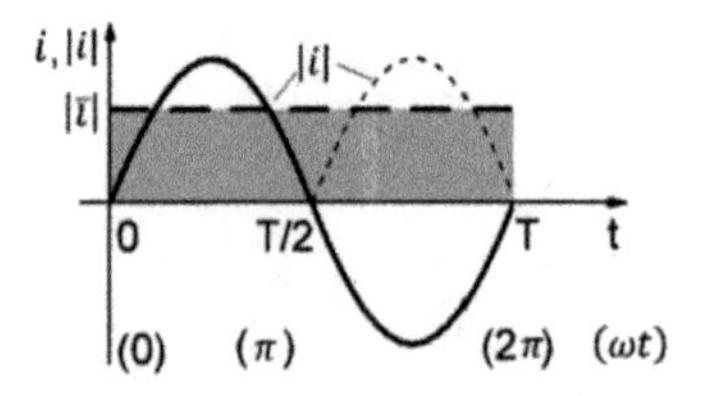

Figure 3.29: Rectified value.

The rectified value of a voltage v(t) or a current i(t) is given by:

$$|\bar{V}| = \frac{1}{T} \cdot \int_0^T |v| \, d(t) \qquad |\bar{I}| = \frac{1}{T} \cdot \int_0^T |i| \, d(t)$$

The rectified values of a **sinusoidal current** and a **sinusoidal voltage** are:

$$|\bar{I}| = \frac{1}{2\pi} \cdot \int_0^{2\pi} |i| d(\omega t)$$

$$|\bar{I}| = \frac{1}{\pi} \cdot \int_0^{\pi} \hat{I} \cdot \sin(\omega t) d(\omega t) = \frac{\hat{I}}{\pi} \cdot (-\cos(\omega t))|_0^{\pi}$$

$$|\bar{I}| = \frac{2}{\pi} \cdot \hat{I}$$

$$|\bar{V}| = \frac{2}{\pi} \cdot \hat{V}$$

The rectified value is of importance in rectifier circuits and when charging accumulators.

3.12 Terms used in power electronics

Rectifiers (AC/DC converters) convert alternating current into direct current.

DC/AC converters convert direct current into alternating or three-phase current.

AC/AC converters convert the alternating current of a voltage V_1 into alternating current of a voltage V_2 with a different voltage level, frequency or phase number.

DC/DC converters convert direct current of a voltage V_1 into direct current of a voltage V_2 with a different voltage level or polarity.

3.13 Charging and discharging of a capacitor

If a discharged capacitor is charged via a resistor (see Figure 3.30) by connecting it to a direct voltage V_0, the voltage applied to the capacitor increases first fast and then ever slower. The voltage connected to the capacitor follows an increasing exponential function:

$$v_C = V_0 \cdot \left(1 - e^{-\frac{t}{\tau}}\right)$$

The charging current $i_C \left(I_0 = \frac{V_0}{R}\right)$, however, is high at first and then decreases according to an exponential function:

$$i_C = I_0 \cdot e^{-\frac{t}{\tau}}$$

τ respectively is the so-called **time constant** and is calculated as follows: $\tau = R \cdot C$

The capacitor is practically charged after the period of time: $t_C = 5 \cdot \tau = 5 \cdot R \cdot C$.

When discharging the capacitor (by flipping the switch in Figure 3.30), the voltage decreases in accordance with a falling exponential function:

$$v_C = V_0 \cdot e^{-\frac{t}{\tau}}$$

Discharging of the capacitor takes five time constants. The voltage and current curves are represented below. The discharging current follows the equation:

$$i_C = -I_0 \cdot e^{-\frac{t}{\tau}}$$

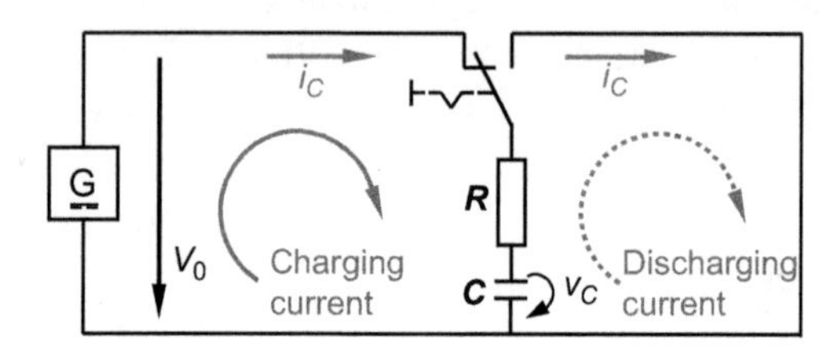

Figure 3.30: Circuit used to charge and discharge a capacitor.

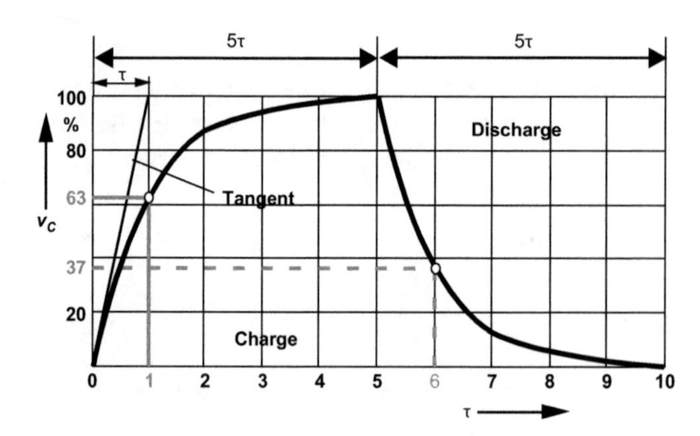

Figure 3.31: Voltage curve of a capacitor during charge and discharge.

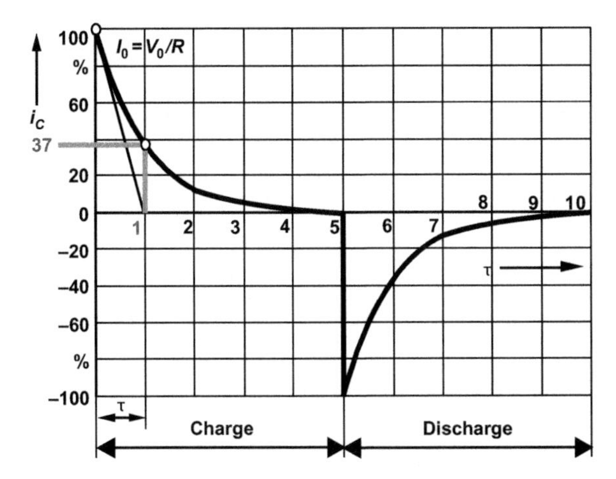

Figure 3.32: Current curve of a capacitor during charge and discharge.

3.14 Review questions

1) Describe the concept of a p-n junction.
2) How does a semiconductor diode work? Circuit symbol, characteristic curve?
3) At which level of voltage does an Si diode start to conduct in forward direction?
4) Zener diode: Circuit symbol, characteristic curve, operating principle
5) How does a limiting circuit work?
6) Resistors: How are resistors labelled and what does the current-voltage characteristic curve of an ohmic resistor look like?
7) What does the resistance-temperature characteristic curve of an NTC resistor or a PTC resistor look like?
8) How does a transistor work? Circuit symbol, designation of terminals?
9) What can a transistor be used for?
10) Sketch and describe the emitter circuit.
11) How does an operational amplifier work?
12) How does a rectifier circuit work and what is the rectified value?

4 The stationary electric field

To separate charges energy is necessary; this energy is stored in the space between the separated charges. The resulting energy space is referred to as electric field. In this field, forces are exerted onto charge carriers. Electrically charged objects are surrounded by an electric field, which represents the state of a certain space (e.g. between electrically charged objects). Said state is characterised by the electric charges that are affected by a force as soon as they enter its space. The electric field strength E measures the force that affects a charged object in an electric field.

The electric field strength at a certain point within the field is defined as the force exerted on a positive point unit charge $Q_+ = 1$ As or C. The mechanical force F as well as the electric field strength E are vector quantities:

$$\vec{E} = \frac{\vec{F}}{Q_+} \qquad [E] = \frac{V}{m}$$

Field types

The pattern of the electric field lines strongly depends on the geometric arrangement of charges. Field lines always enter or exit the charge carrier vertically.

Radial symmetric field

Homogeneous field

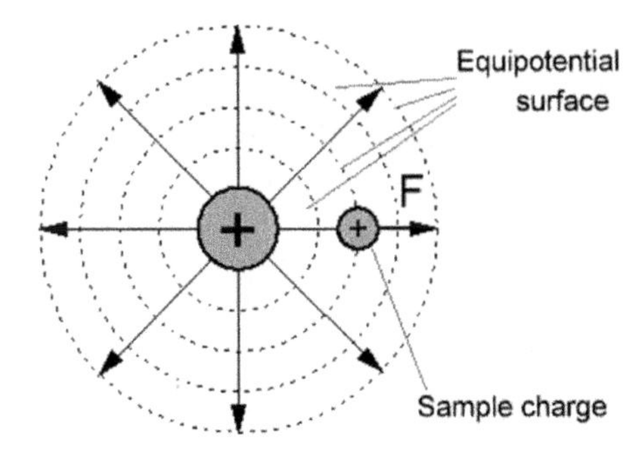

Figure 4.1: Field line path of a charged sphere.

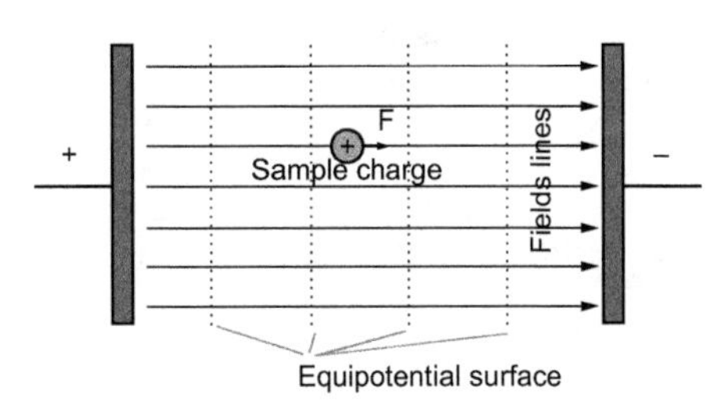

Figure 4.2: Field line path between two charged plates.

https://doi.org/10.1515/9783110521115-004

Displacement work

The force F depends on the field strength E and the quantity of electric charge Q.

Energy level

If a charge is moved using work against the field force, it subsequently has a higher energy level.

Electric potential

To obtain information on the possible (work) potential of an electric field, the potential must relate to the charge (potential per unit charge). The reference potential can be determined arbitrarily. Usually, the negative electrode is defined as zero potential. A surface with same electric potential at every point is called an equipotential surface.

Voltage

When energy is gained or lost, a charge is transferred from an electric potential 1 to another potential 2. The potential difference is called voltage V (corresponds to the striving for balance of separated charges). Voltage is also referred to as potential difference.

Energy equivalents

Converted energy changes its form but not its value. In conductive materials, charge carriers continuously perform irregular movements. If a movement in a certain direction additionally superposes the irregular movements, this is called **electric current**. Current flow is only possible in a closed cycle, referred to as **electric circuit**. Furthermore, a **voltage source** is required to cause the current to flow. A simple circuit consists of a voltage source, a lightbulb and the connecting components. The current flowing in such a circuit is called **direct current**. The electric current flow is referred to as **electric current intensity** or just **current**.

4.1 Electric current I

$$I = \frac{dQ}{dt} \qquad [I] = 1\frac{C}{S} = 1A$$

At the constant current I, the electric charge Q moving through the cross-section of a conductor during time t equals:

$$Q = I \cdot t \qquad [Q] = As = C$$

One coulomb is hence the electric charge that flows through the cross-section of the conductor per second at a current of one ampere. The lowest negative electric charge possible is that of an electron. It is referred to as elementary charge and has the value $e = 1.602 \cdot 10^{-19}$ C. The proton's charge is positive but equal in value. The direction of movement of current was determined as **sense of direction of the current flow**. Conventionally, it is the direction of movement of *positive* charge carriers. The negative electrons move against the sense of direction of the current flow.

Current density J:

$$J = \frac{I}{A} \qquad [J] = \frac{A}{m^2}$$

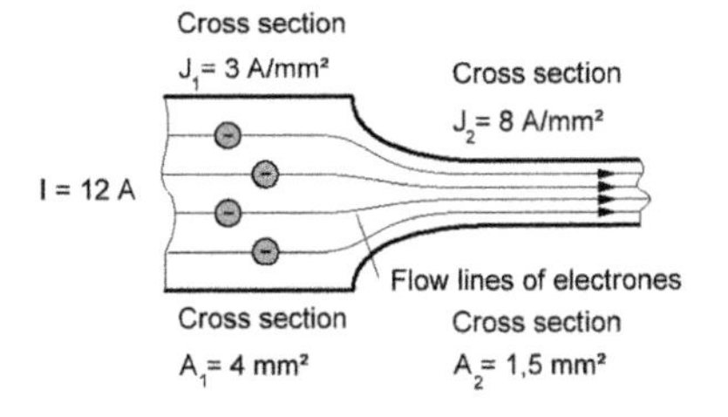

Figure 4.3: Visualisation of the quantity current density.

Drift velocity of electrons

$$v = \frac{I}{e \cdot n \cdot A}$$

v Medium drift velocity of electrons in $\frac{m}{s}$
n Electron density in m^{-3}
A Flow cross-section in m^2

4.2 Electric potential φ

There is no work performed on a charge if it is moved in the electric field of another charge. This property

$$W = q \cdot \oint_1 \vec{E} \cdot d\vec{s} = 0$$

applies to all stationary electric fields. According to mathematics, all vector fields which obey the above integral relation can be represented by the gradient of a scalar function of the local coordinates $\varphi(x, y, z)$. This leads to:

$$\varphi = \varphi(x, y, z)$$

$$E = -grad\ (\varphi) = -\left(\frac{\partial\varphi}{\partial x}\cdot\vec{e_x} + \frac{\partial\varphi}{\partial y}\cdot\vec{e_y} + \frac{\partial\varphi}{\partial z}\cdot\vec{e_z}\right) = -\frac{d\varphi(x, y, z)}{d\vec{s}} \qquad \text{(Equation A)}$$

$$[\varphi] = 1V = 1\frac{Nm}{As}$$

The scalar local function $\varphi(x, y, z)$ is called electric potential with the unit V or $\frac{Nm}{As}$. The negative sign is determined arbitrarily. Calculating $\varphi(x, y, z)$ using the electric field strength, we obtain:

$$\int_1^2 E\cdot d\vec{s} = -\int_1^2 \frac{d\varphi}{d\vec{s}}\cdot d\vec{s} = -\varphi_2 + \varphi_1$$

The left integral is the voltage between points 1 and 2. The potential difference is hence proportional to the work used to move a charge in the electric field along a path s.

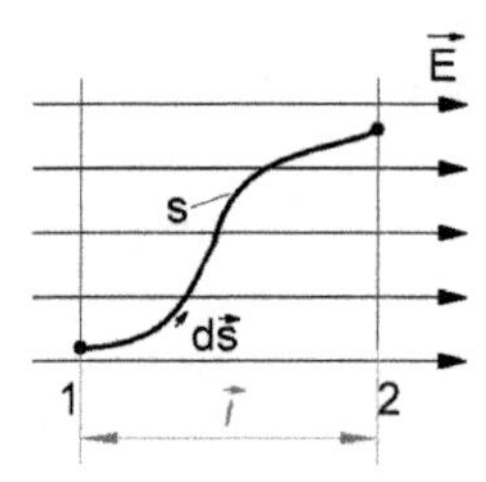

Figure 4.4: Moving a charge in the electric field along a path s.

The integral allows to arbitrarily choose the potential of either point 1 or 2, as only a statement regarding the difference is made. This enables us to use simplifications when calculating electric fields. Presuming $\vec{E}$ is constant, we obtain for the above array (independent of the distance):

$$V_{12} = \int_1^2 E\cdot d\vec{s} = E\cdot l = -\varphi_2 + \varphi_1$$

If φ_2 is arbitrarily set to zero, we obtain:

$$V_{12} = E\cdot l = 0 + \varphi_1$$

We notice that with positive voltage the potential of $\varphi_1 = E\cdot l$ decreases towards $\varphi_2 = 0$.

A positively charged particle is accelerated in the direction of the decreasing potential. Therefore, the choice of sign in the Equation A (page 73) seems plausible, because it would contradict our conception if particles were accelerated in the direction of higher potentials.

A potential can also be assigned to two points in a circuit. The current is conventionally defined as the flow of positive charge carriers (passive sign convention, PSC) and flows from the higher (+) to the lower (–) potential.

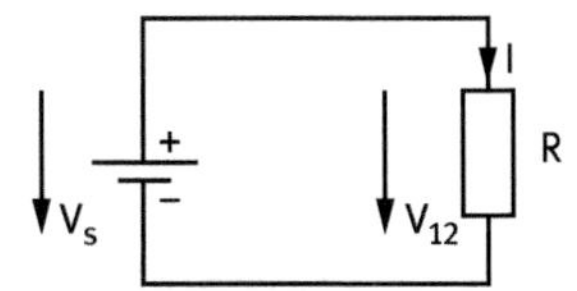

Figure 4.5: Potentials of a circuit.

Therefore, the term "voltage drop" for $V_{12} = \varphi_1 - \varphi_2$ seems reasonable. Additionally to the field lines, lines can be inserted into a field pattern, which connect points of constant potential. These lines are called equipotential lines.

For the field of a point charge, the potential results from

$$\int \vec{E} \cdot d\vec{s} = \int \frac{q}{4 \cdot \pi \cdot \varepsilon_0 \cdot \varepsilon_r \cdot r^2} \vec{e_r} \cdot \vec{e_r} \cdot dr = -\frac{q}{4 \cdot \pi \cdot \varepsilon_0 \cdot \varepsilon_r \cdot r^2} + C = -\varphi$$

$$\varphi(r) = \frac{q}{4 \cdot \pi \cdot \varepsilon_0 \cdot \varepsilon_r \cdot r^2} + C$$

As φ (r) only depends on the radius r, all points with the radius r are constant in potential.

If one looks at a point charge in space, all points with constant potential form a spherical surface (=equipotential surface). The electric field lines are always perpendicular to the equipotential surfaces. This results from the definition of the gradient.

In a constant homogeneous electric field, the following applies:

$$\int_1^2 \vec{E} \cdot d\vec{s} = \int_1^2 \vec{E} \cdot \vec{e_x}\, dx = [E \cdot x + C]_1^2 = -\varphi_2 + \varphi_1$$

If a potential is arbitrarily assigned to an equipotential surface ($x = 0 : \varphi_1 = 0$, $C = 0$), the potential φ_2 is clearly defined, depending on the distance x.

$$\varphi_2(x) = -E \cdot x$$

From the field pattern one can deduce qualitative information about changes in value of the field strength. A small distance between adjacent field lines (high field line density) indicates high field strength.

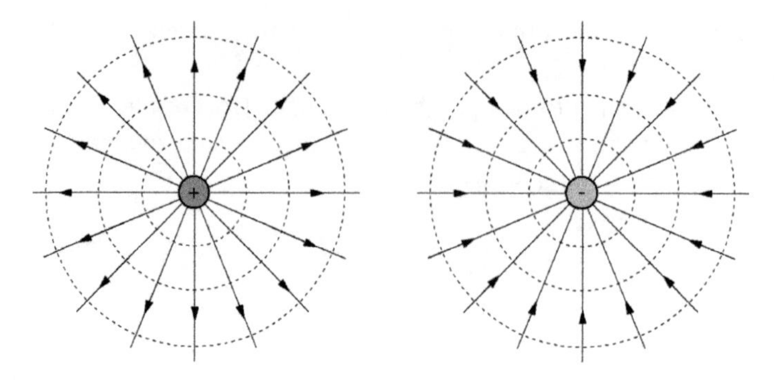

Figure 4.6: Electric field lines and equipotential lines (dashed) of a positive and negative point charge.

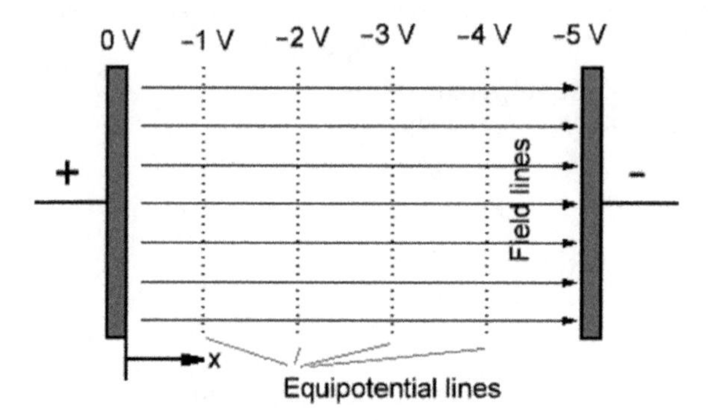

Figure 4.7: Potential lines of a homogeneous electric field (e.g. plate capacitor).

4.3 Electrostatic induction

If two metal plates carry the electric charges +Q and -Q, an electric field exists between them. When a metallic conductor is brought into the field, electric forces act on its charges. The mobile electrons are displaced opposite to the field direction whereby a charge separation takes place.

The charge redistribution in an originally neutral object is called electrostatic induction. Through charge separation another electric field forms within the conductor, which counteracts the outer field. The charge redistribution is completed when the opposing field is equal to the outer field. The inside of the conductor is thus field free meaning that this space is shielded against the outer field (Faraday cage – charges on the metal surface).

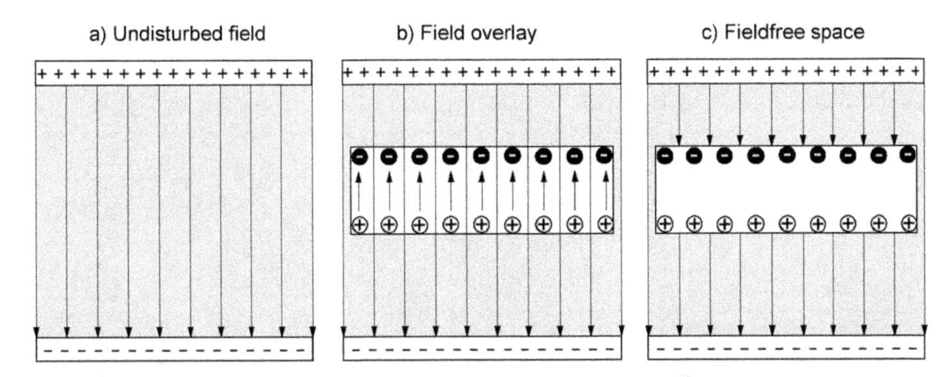

Figure 4.8: Electrostatic induction is the redistribution of electric charges in a conductor under the influence of an electric field.

4.4 Polarisation

The redistribution of charges in the electric field does not only take place in electric conductors but also in electrically insulating materials. Due to the charge carriers being stationary in insulators, the electric charges can only be moved or twisted within their associated atom or molecule. Insulating materials have hardly any free electrons; charges are only moved slightly (weak induced voltage).

Therefore, due to the impact of the electric field small, electric dipoles form on the previously neutral insulating material atoms. If charges are electrostatically separated from an originally neutral object, we talk about polarisation.

There are two types of polarisation:

- **Dielectric polarisation** occurs with certain insulating materials. Within the molecules, charges are aligned. These molecules are called dipoles.
- **Paraelectric polarisation**: Insulating materials are already organised as dipoles. However, without an outer electric field, the electric effects cancel each other out due to the disordered thermal movement (a). If an outer field is applied, the dipoles align themselves according to their polarities (b).

4.5 The electric displacement flux Ψ

Electric fields are the sum of all field lines. The fields occur due to the redistribution of charges; the sum of all field lines therefore is called displacement flux Ψ, which corresponds to the amount of separated charges. The following applies: $\Psi = Q$.

The displacement flux is the description of the electric charge with regard to the field. The number of field lines per perpendicular surface unit varies depending on the development of the field. The displacement flux per perpendicular surface unit is called displacement flux density D. The following applies: $\vec{D} = \frac{\Psi}{A}$. It is a vector and behaves proportionally to the electric field strength. In a vacuum, the relation: $D = \varepsilon_0 \cdot E$ applies, in other substances, it is $D = \varepsilon_0 \cdot \varepsilon_r \cdot E$. Thereby, ε_0 is the electric constant and ε_r the material-dependent permittivity.

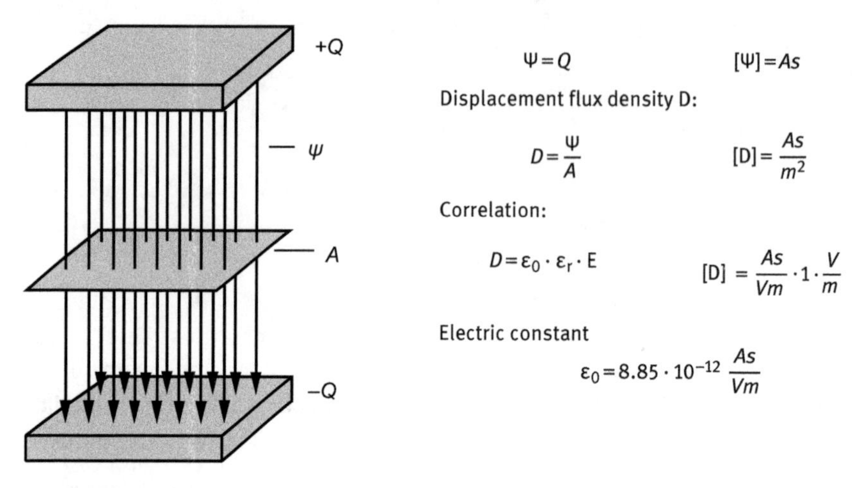

Figure 4.9: The electric displacement flux Ψ is the quantity of electric charge Q redistributed by an electric field.

4.6 Dielectric

Electrically insulating materials in components where **strong electric fields** occur are usually called dielectrics (e.g. cable insulation, capacitors). The **dielectric strength E_d** of a dielectric is the maximum permissible electric field strength at which the material is not destroyed. It is indicated in $\frac{kV}{mm}$.

4.6.1 Permittivity ε (formerly known as dielectric constant)

Permittivity (permeability of a material for electric fields)

Table 4.1: Selection of technically used dielectrics.

Material	ε_r	E_d *in kV/mm*
Air (normal pressure)	1	2.1
Water (distilled)	80	–
Natural mica	6...8	30...70
China	5...6	35
Polyethylene (PE)	2.3	60...90
Polystyrene (PS)	2.3...4.2	35
Epoxy resin	3.7...4.2	35
Silicone rubber	2.5	20...30

4.7 The capacitor

Capacitors in their basic form (see Figure 4.10) consist of two spatially extended electrodes insulated from each other (e.g. plates, cylinders, spheres). When applying the voltage V, the capacitor gains the electric charge Q and stores energy W in its electric field.

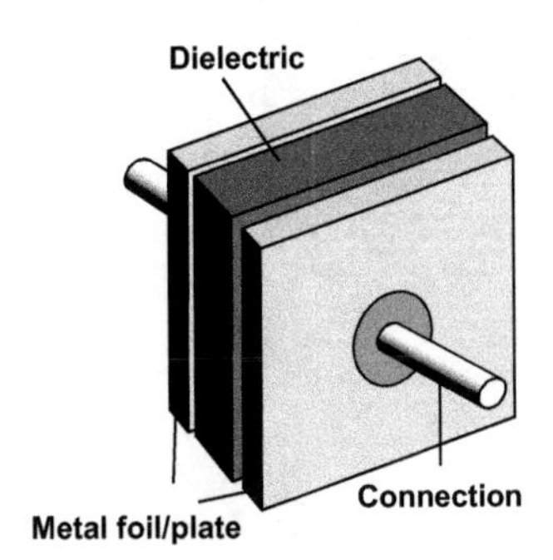

Figure 4.10: Plate capacitor.

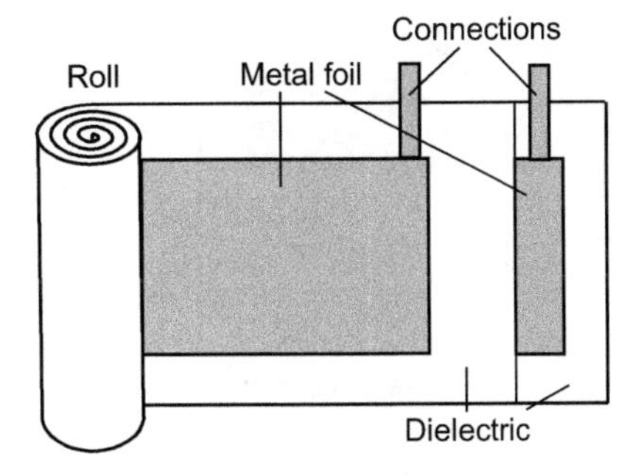

Figure 4.11: Structure of a film capacitor.

Essentially, there are three main groups of capacitors:
- **Ceramic capacitors:** Multilayer ceramic chip capacitors (MLCCs) are the most common construction type of ceramic capacitors according to quantity. They consist of stacked metallised ceramic layers that serve as carriers and are contacted with the terminal surfaces at their front surfaces.
- **Film capacitors**: They consist of two strips of metal foil with a thin paper or plastic film (see Figure 4.11). The entire structure is wrapped with additional insulating film and rolled up into a "roll". With large capacitors sometimes several such rolls are connected in parallel and are in an oil bath for reasons of insulation.
- **Electrolytic capacitors** (see Figure 4.12): A thin oxide layer that acts as dielectric forms through electrolysis on one electrode (anode, mostly out of aluminium sheet or foil).[43] The other electrode (cathode) is an electrolyte that is absorbed into a paper and only is in contact with one metal strip serving solely as supply. The array is also mostly depicted as a film roll. Such capacitors have a great storage capacity while they can be produced relatively cheaply due to their dielectric having a thin coating (oxide layer). However, the operating voltage must not exceed 500 V and the electric field may only be applied in direction of the creation of the oxide layer,[44] otherwise the oxide layer is destroyed.

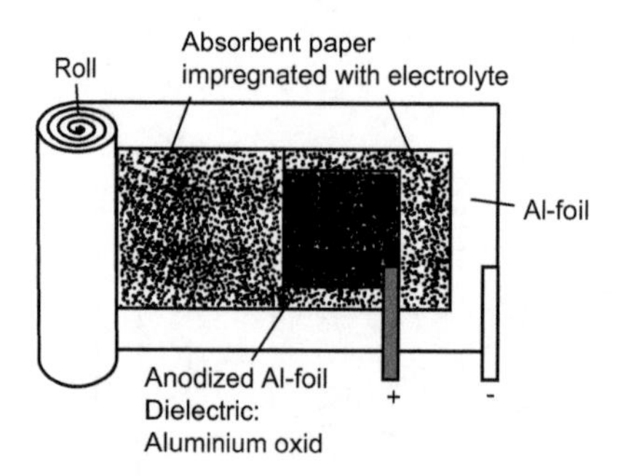

Figure 4.12: Structure of an electrolytic capacitor.

Figure 4.13: Structure of a multilayer ceramic chip capacitor.

43 The oxide layer is only a few thousandths of a millimetre.
44 The thin oxide layer is decomposed with the wrong polarity. After applying the operating voltage, the capacitor is heated rapidly to a very high temperature and consequently destroyed (explosion!).

4.7.1 Electrical capacitance C

We can determine in experiments that the charge Q of a plate capacitor increases in a linear way with the applied voltage: $Q \sim V$

More precisely:

$$Q = C \cdot V$$

C is the coefficient of proportionality and is called "capacitance".

$$\mathbf{C = \frac{Q}{V}}$$

$$[C] = \frac{[Q]}{[V]} = 1\frac{As}{V} = 1\ Farad = 1\ F$$

Each arbitrary array of two conductors, separated by a dielectric, has a capacitance. Possible capacitance range: $10^{-12} \ldots \approx 1\ F$. Recent developments of so-called super-capacitors are based on special chemical processes and allow capacitances of several 1,000 F, however with relatively small operating voltages.

The capacitance of a capacitor depends on the geometry and the primitiveness of the dielectric. An analogous correlation also applies to the electric field strength E. For a plate capacitor with the plate distance d and the plate area A the following applies:

$$C = \varepsilon_0 \cdot \varepsilon_r \cdot \frac{A}{d} \qquad E = \frac{V}{d}$$

C Capacitance in F
d Distance between the plates in m
A Surface area of the plates in m^2
E Electric field strength in $\frac{V}{m}$

4.7.2 Series connection of capacitances

In series connections, the same charging current flows through all capacitors as soon as a voltage is applied. Based on this, the following regularities can be deduced:

- In a series connection, all capacitors have the same electric charge.
- The partial voltages behave towards one another in the reverse way as the related partial capacitances do.
- The reciprocal of the total capacitance equals the sum of reciprocals of the partial capacitances.

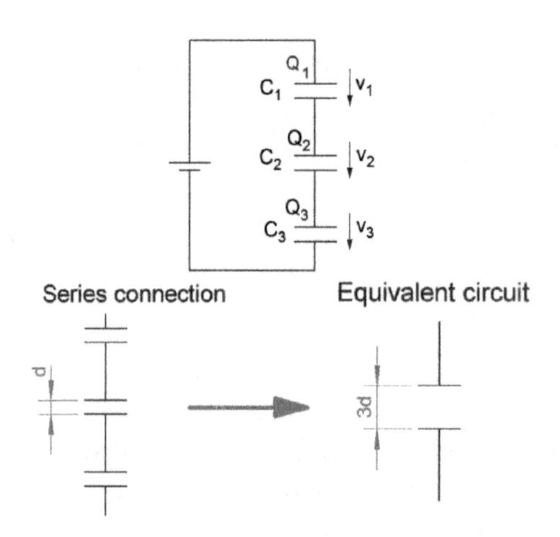

$$Q_1 = Q_2 = Q_3$$

$$\frac{V_1}{V_2} = \frac{C_2}{C_1}$$

$$\frac{1}{C} = \frac{1}{C_1} + \frac{1}{C_2} + \frac{1}{C_3}$$

Therefore, the following generally applies:

$$\frac{1}{C_t} = \sum_i \frac{1}{C_i}$$

For the series connection of 2 capacitors, the following applies:

$$C = \frac{C_1 \cdot C_2}{C_1 + C_2}$$

Figure 4.14: Series connection of capacitances.

4.7.3 Parallel connection of capacitances

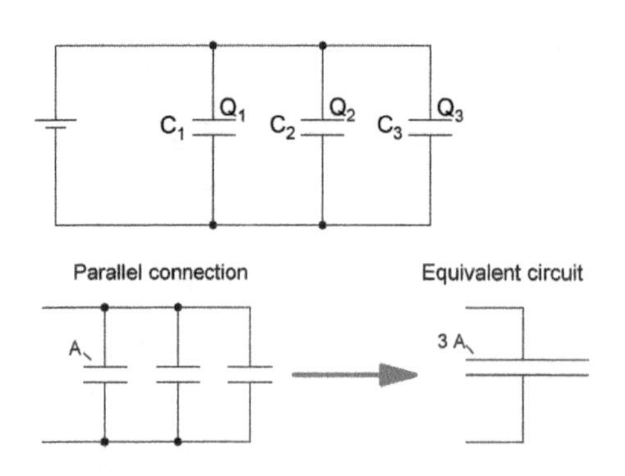

$$Q = Q_1 + Q_2 + Q_3$$

$$\frac{Q_1}{Q_2} = \frac{C_1}{C_2}$$

$$C = C_1 + C_2 + C_3$$

Therefore, the following generally applies:

$$C_t = \sum_i C_i$$

Figure 4.15: Parallel connection of capacitances.

The same voltage applies to all components when capacitors are connected in parallel; this is also true for the parallel connection of ohmic resistors. Following this principle 3 laws can be deduced:

- In parallel connection the overall charge equals the sum of the partial charges.

- The partial charges and thus also the partial currents behave like partial capacitances.
- The total capacitance equals the sum of the partial capacitances.

4.8 Energy content of an electric field W_{el}

A capacitor is connected to a direct voltage through a resistor R. Current flows until the capacitor is charged to the voltage V. Now the capacitor has the charge Q and the voltage V. The following relation applies $Q = C \cdot V$, i.e. the charge Q is proportional to the voltage V (with coefficient of proportionality C). The area in Figure 4.16 corresponds to energy W of the charged capacitor. This triangle has the area $\frac{1}{2} \cdot Q \cdot V$. Instead of Q we use $C \cdot V$. The electrical energy is hence calculated as follows:

$$W_{el} = \frac{1}{2} \cdot C \cdot V^2$$

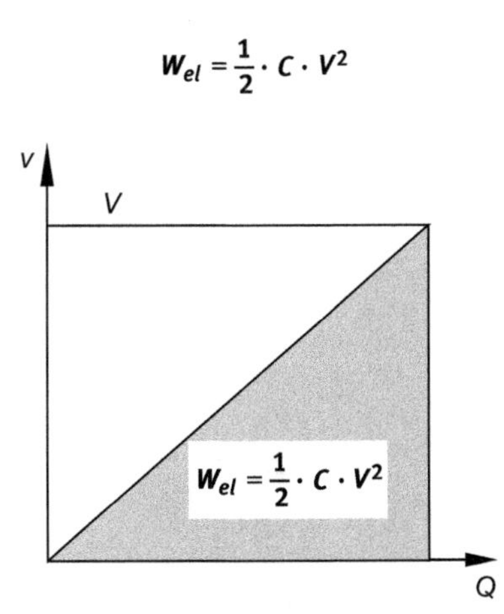

Figure 4.16: V versus Q; deriving the energy content of an electric field.

4.9 Review questions

?

1) Explain the terms electrostatic induction and polarisation.
2) What is the definition of capacitance C and what are the rules that apply to series connections and parallel connections of capacitances?
3) What influence does permittivity have on the capacitance of a capacitor?
4) In what way does the distance of the plates d influence the capacitance of a plate capacitor?
5) Describe the processes during charging and discharging (voltage and current curve) of a capacitor.

4.10 Exercises

EXERCISE 4.1

Calculate the capacitance C of two parallel rectangular plates in vacuum with an area of 100 mm^2 and a distance on 10 mm. How changes the capacitance when polyethylene ($\varepsilon_r = 2.4$) is used as dielectric medium?

EXERCISE 4.2

Three capacitors with $C_1 = 68$ µF, $C_2 = 150$ µF and $C_3 = 40$ µF were connected parallel. Which value has the total capacitance C_p?

EXERCISE 4.3

Three capacitors with $C_1 = 68$ µF, $C_2 = 150$ µF and $C_3 = 40$ µF were connected in series. Which value has the total capacitance C_s?

EXERCISE 4.4

A capacitor C = 220 µF is fed by a voltage V of 100 V until it is fully loaded. Which electric energy E is stored in the capacitor?

5 The magnetic field

5.1 The term "field"

Magnetism[45] is a **physical phenomenon** that manifests itself as a force between magnets, magnetised or magnetisable objects and mobile electric charges, like e.g. current-carrying conductors. This force is conveyed through a magnetic field[46] (vector field[47]) that, on the one hand, is created by these objects and, on the other hand, affects them. Magnetic fields occur with any movement of electric charges. A "field" is generally defined as a space where physical laws apply to certain circumstances.

In **permanent magnets**, magnetism is caused by Ampère's molecular currents (electrons rotating around the nucleus create a very small spin current and electrons rotate around themselves – electron spin). In permanent magnets the magnetic effects do not cancel each other out. Demagnetising them requires a considerable amount of energy.

If a **magnetic field spreads in a material body**, the magnetic properties of the substance influence the intensity of the field. The flux density B does not display the same field strength H as in a vacuum. This is due to the atomic structure of the substances. The electrons rotating around their own axis (electron spin) and the nucleus generate spin currents which create magnetic fields perpendicular to the circular orbit (elementary fields). The elementary fields usually cancel each other out without an additional external magnetic field.

The magnetic and electric **expansion** happens at the speed of light and is one of the properties of space. Not only space filled with matter, but also empty space has physical properties.

- The **field strength** is the force (amount and direction) that the field exerts on the **standard body**: the vector field. The field strength is a vector.
- Field lines are used to visually describe a field; they are only a mental tool and not a physical reality.
- Magnetic fields arise from mobile charges. The field lines **encircle the current.**
- The magnetic field is a **source-free vortex field,** it only consists of vortices und **has neither sources nor sinks.**
- The magnetic field lines are **self-contained**; they come out of the north pole and go into the south pole.

45 The name "magnet" probably derives from the Greek place name Magnesia, where permanent magnets were sold as a cure to various physical ailments.

46 "Field" generally refers to a space in which states act according to physical laws.

47 In a vector field, each point in a space has its respective vector.

https://doi.org/10.1515/9783110521115-005

5.1.1 Right-hand screw rule

! The magnetic field lines have a direction and progress clockwise if you look at the conductor in current direction. As a mnemonic aid you can think of a screw (with a right-handed thread) that is being screwed into the conductor in direction of the current. The direction of rotation of the screw indicates the direction of rotation of the field lines. (=right-hand screw rule)

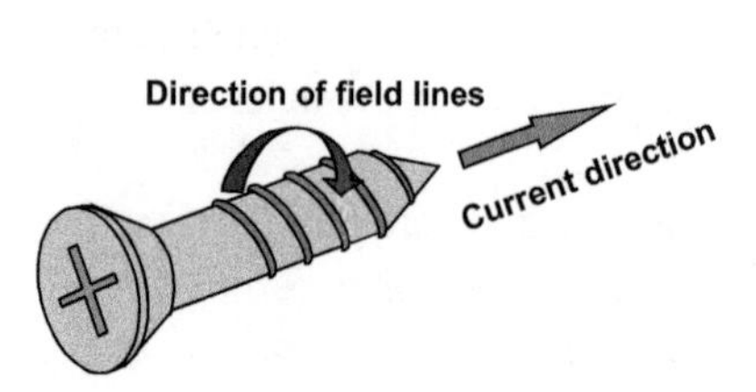

Figure 5.1: Right-hand screw rule.

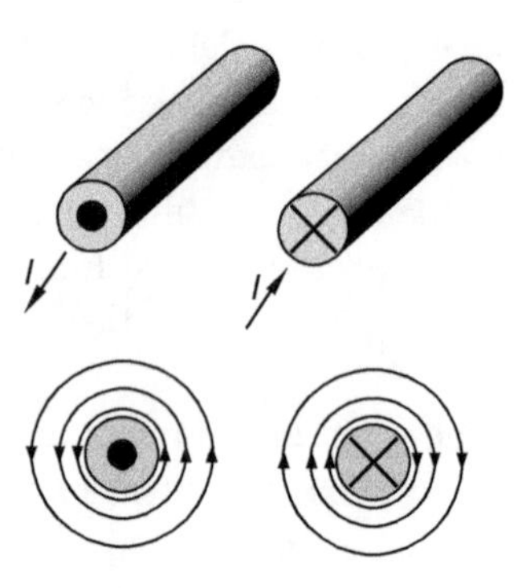

Figure 5.2: Direction of magnetic field lines with different current directions.[48]

5.1.2 Coil rule (right-hand coil rule)

Using the coil rule (see Figure 5.3) we can determine the north and south pole of a current-carrying coil.

Place the right hand around the coil so that the fingers point in current direction; your splayed thumb will point to the north pole of the coil.

48 The current direction into the conductor is represented by an X, the current direction out of the conductor is represented by a dot.

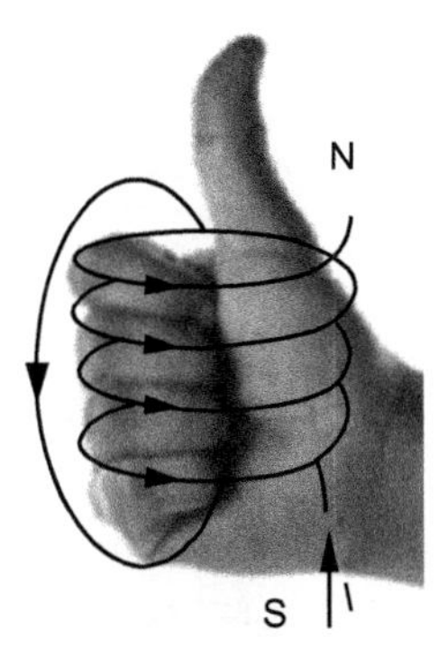

Figure 5.3: Direction of magnetic field lines with different current directions.

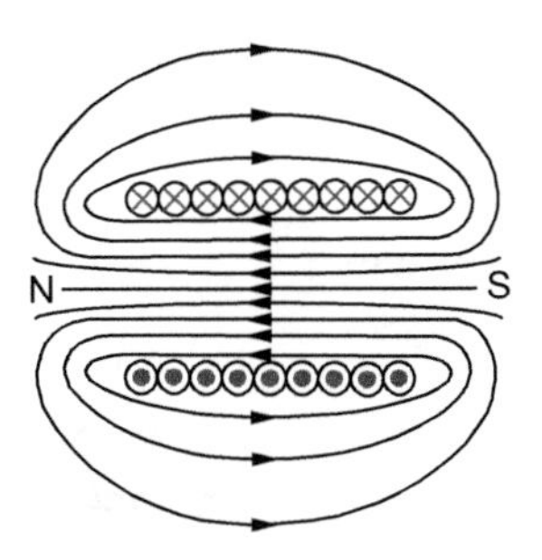

Figure 5.4: Magnetic field of a coil.

5.2 Magnetic field quantities

5.2.1 The magnetomotive force *MMF*

The total of all magnetic field lines, the so-called magnetic flux, is caused by the electric current. Thereby, it is important to know if the current only flows through one winding or is used several times. The magnetic effect is proportional to the current I and to the number of windings N. The magnetomotive force MMF is therefore defined as:

$$MMF = I \cdot N$$

MMF Magnetomotive force in A
I Current in A
N Winding number

5.2.2 The magnetic field strength *H*

The concentration of magnetic energy is crucial for the magnetic effect. The greater the magnetomotive force MMF and the shorter the coil length l_m, the greater the magnetic effect, which corresponds to the mean field-line length.

Hence, the magnetic field strength H is defined as follows:

$$H = \frac{MMF}{l_m} = \frac{I \cdot N}{l_m}$$

H Magnetic field strength in $\frac{A}{m}$
MMF Magnetomotive force in A

I Current in A
N Winding number
l_m Coil length in m

5.2.3 The magnetic flux ϕ

The total number of magnetic field lines of a permanent magnet or current-carrying coil is called magnetic flux ϕ. The term was coined based on the electric current flow I. The magnetic flux ϕ can only be measured by its impacts. The unit of magnetic flux is the volt second (Vs), also specifically referred to as weber (Wb).

5.2.4 The magnetic flux density B

The force of the magnet is higher, the denser the magnetic field lines are, meaning the higher the magnetic flux ϕ and the smaller the area A penetrated by it.

The magnetic flux density is defined as:

$$B = \frac{\Phi}{A}$$

The **magnetic flux density *B*** (or magnetic induction) uses the unit Tesla[49] (T). It is proportionality constant and describes the strength of the magnetic field (which surrounds e.g. a current-carrying conductor).

Figure 5.1 lists examples of flux densities occurring in everyday life to get a better idea of the magnitude of this unit.

Table 5.1: Examples of different flux densities.

	Magnetic flux density *B*
Earth's magnetic field	40 μT
Magnet resonance imaging (MRI)	3 T
Max. value for electric appliances in the household	100 μT
Electric drill (distance of 30 cm)	≈ 3 μT

49 Nikola Tesla, Croatian physicist, *1856 †1943. He was awarded the honorary doctorate of the Graz University of Technology (TU Graz) in 1937. To honour his legacy, the testing hall at the Institute for High Voltage Engineering at the TU Graz was renamed "Nikola Tesla Laboratory" on the occasion of Tesla's 150th birthday.

5.2.5 The permeability μ

The magnetic flux density B and the magnetic field strength H are interconnected:

$$B = \mu \cdot H$$

B Magnetic flux density in T
H Magnetic field strength in $\frac{A}{m}$
μ Absolute permeability in $\frac{Vs}{Am}$

The quantity μ is called permeability (or magnetic conductivity) and determines the permeability of matter in magnetic fields. The relative permeability μ_r indicates the factor, by which the respective substance magnetically conducts better (or worse) than vacuum; the permeability of vacuum is designated by the magnetic constant μ_0. The permeability μ of any substance can be calculated through the product:

$$\mu = \mu_0 \cdot \mu_r$$

μ_0 Magnetic constant (vacuum permeability) in $\frac{Vs}{Am}$ $\qquad \mu_0 = 4\pi \cdot 10^{-7} \frac{Vs}{Am}$
μ_r Relative permeability $[\mu_r] = 1$

In paramagnetic and diamagnetic substances, the relative permeability μ_r is independent from the field strength H; the magnetisation curve is a straight line through the origin. The following applies:

$$\mu = \frac{B}{H}$$

Substances are separated into three groups according to their permeability (Table 5.2).
- **Ferromagnetic substances** (iron, nickel, cobalt): They magnetise in an external magnetic field in a way. Their internal magnetic flux density highly increases compared to the external space. Hence the field lines inside the substance are closer together.[50] Important for many technical applications like generators.
- **Paramagnetic substances** (air, Al, Pt): Paramagnets have a magnetisation unequal to zero only as long as they are in an external magnetic field.

50 Hence, they are good magnetic conductors.

- **Diamagnetic substances** (bismuth, carbon, water, glass, gold, lead): Their internal magnetic field slightly decreases proportional to the strength of the applied magnetic field.

Table 5.2: Relative permeability.

Ferromagnetic substances		Paramagnetic substances		Diamagnetic substances	
Iron, unalloyed	< 6,000	Air	1.0000004	Mercury	0.999975
Electric sheet steel	> 6,500	Oxygen	1.0000003	Silver	0.999981
Ion nickel alloys	< 300,000	Aluminium	1.000022	Zinc	0.999988
Soft magnetic ferrites	> 10,000	Platinum	1.000360	Water	0.899992

Diamagnetism and paramagnetism are atomic properties (both are considered non-magnetic in a technical sense).

Ferromagnetism

The properties of ferromagnetic substances are due to their particular crystalline structure. In the structure, small crystalline domains ("Weiss domains") are formed that act like small permanent magnets. The Weiss domains can be oriented or turned through external magnetic impact. The orientation happens abruptly and can be made audible by means of an induction coil and an amplifier. By the alignment of the Weiss domains, the length of the magnet changes; i.e. constant change in the orientation of magnetisation leads to a constant change in length of the magnet → "**magnetostriction**"[51] (100 Hz humming sound coming from transformers). All ferromagnetic substances lose their ferromagnetic properties above the **Curie temperature** and become paramagnetic (e.g. for iron 769 °C).

Ferromagnetism is a crystalline property. When brought close to a magnet in a magnetic field, a ferromagnetic substance turns into a magnet itself without reducing the strength of the source magnet. In ferromagnetic substances, the **relative permeability μ_r** is significantly higher than 1 and (non-linearly) dependent from the source **field strength H**. Hence, for the permeability μ of ferromagnetic substances, the following applies:

51 Magnetostriction is the deformation of magnetic (especially ferromagnetic) substances as a result of an applied magnetic field. Thereby the object experiences an elastic deformation in length at constant volume.

$$\mu = \frac{\Delta B}{\Delta H} \quad \text{precisely:} \quad \mu = \frac{\mathrm{d}B}{\mathrm{d}H}$$

The magnetic flux density *B(H)* is deduced in practice directly from the magnetisation curve (see Figure 5.5).

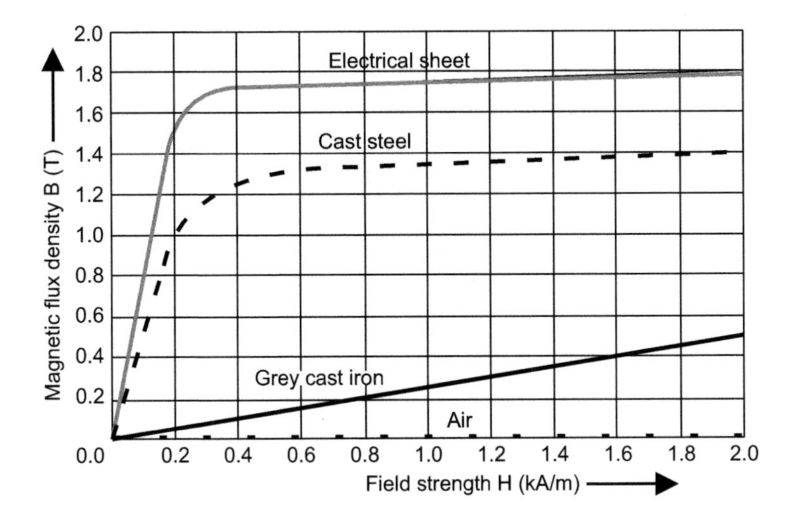

Figure 5.5: Magnetisation curve of selected substances.

Hysteresis loop

The magnetisation of ferromagnetic substances usually does not depend on the external magnetic field linearly. It is possible, to magnetise ferromagnetic materials up to the point of saturation (flat part of the curve). Furthermore, the magnetisation depends on the prior magnetisation, which is referred to as magnetic memory. This behaviour is described by means of a hysteresis loop (see Figure 5.6). Increasing the magnetic field strength H for the first time,[52] the curve *B(H)* follows the dashed curve (**initial curve**). If the magnetic field strength is set to zero again, the course of $B(H)$ follows the upper curve; the flux density B_R (**remanent flux density**) remains.[53] In order to return the magnetic flux density B to zero, the magnetic field strength needs to be polarised reversely and increased up to the value H_C (**coercivity**). From $-H$ to $+H$ the course of $B(H)$ follows the lower curve.

52 . . . if there is no remanence.

53 The ferromagnetic substance has turned into a permanent magnet.

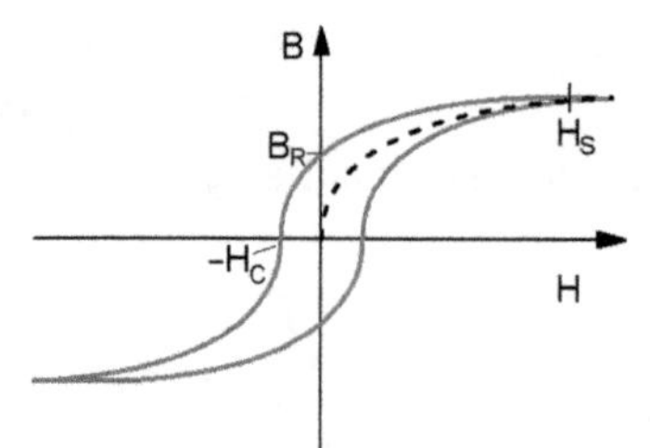

Figure 5.6: Hysteresis loop.

The hysteresis loop looks different for the various types of ferromagnetic materials. A broad hysteresis loop indicates **hard magnetic** materials. If the hysteresis loop is narrow, we talk about **soft magnetic** materials. For electric sheet metal in machines, an ideal soft magnet (hysteresis loop would be a line) is desirable[54] while for the magnetic storage of bits an ideal hard magnet is preferred.[55]

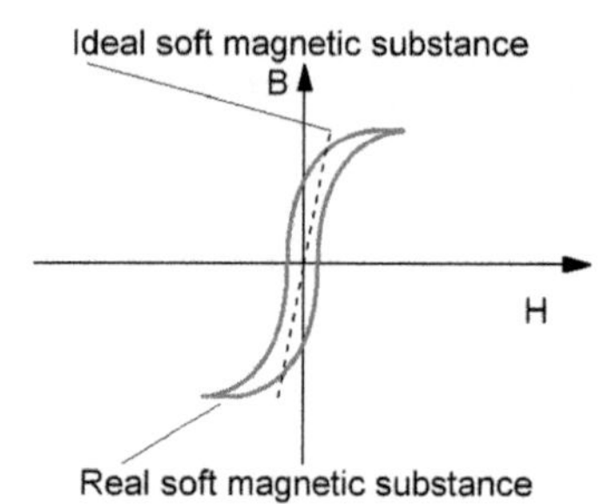

Figure 5.7: Hysteresis loop of soft magnetic substances.

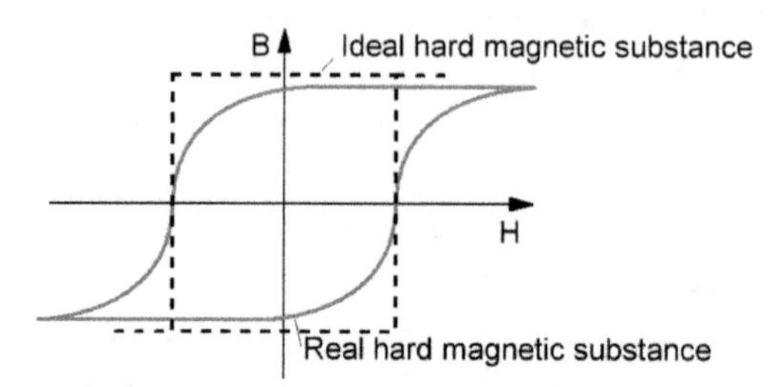

Figure 5.8: Hysteresis loop of hard magnetic substances.

54 ... as with ideal soft magnets, hysteresis losses (energy that is used for the change of alignment of the Weiss domains) would not occur.

55 When changing the field direction, the magnetisation turns all of a sudden – ideal to illustrate the states "0" and "1". No information can be stored in an ideal soft magnetic material as no remanent flux density remains.

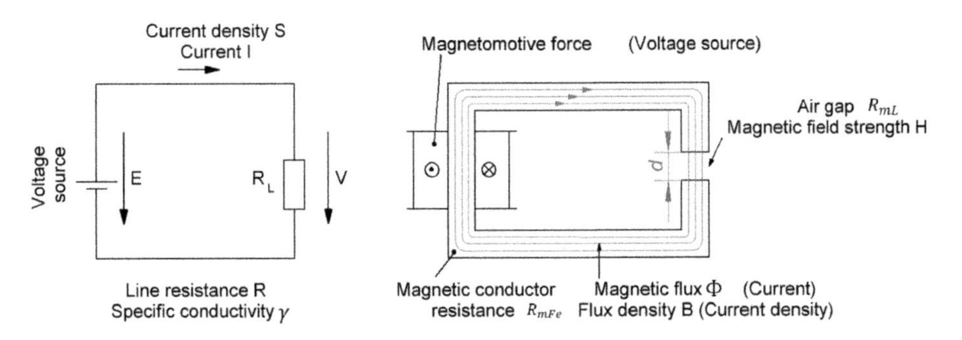

Figure 5.9: Comparison of functional principles.

5.2.6 Comparison of electric and magnetic circuit

In the electric circuit, the voltage causes the current; the electrical resistance works against the generation of current (flow). In the magnetic circuit, the magnetomotive force causes the magnetic flux ϕ; the magnetic resistance R_m hinders the generation of the magnetic flux ϕ. Table 5.3 provides a comparison of the respective quantities.

Table 5.3: Parameter comparison electric and magnetic circuit.

Electric circuit		Magnetic circuit	
Battery; electric voltage source creates *EMF*	V	Coil; magnetic voltage source generates *MMF*	A
Electric current I	A	Magnetic flux ϕ, field lines	Vs = Wb
Current density S	$\frac{A}{m^2}$	Flux density (Induction) B	$\frac{Vs}{m^2} = T$
Electric Field Strength E	$\frac{V}{m}$	Magnetic field strength H	$\frac{A}{m}$
Load resistance R_L	Ω	Air gap magnetic resistance R_{mL}	$\frac{1}{H}$
Line resistance R	Ω	Magnetic conductor resistance R_{mFe} of iron	$\frac{1}{H}$
Specific conductivity of the conductor material γ	$\frac{S}{m}$	Specific magnetic conductivity μ of iron	$\frac{H}{m}$

Ohm's law for magnetic circuits:

$$MMF = R_m \cdot \phi$$

MMF	Magnetomotive force in A (56)
ϕ	Magnetic flux in Wb or Vs
R_m	Magnetic resistance in $\frac{1}{\Omega s}$

Magnetic resistance R_m:

$$R_m = \frac{l}{\mu \cdot A}$$

R_m	Magnetic resistance in $\frac{1}{(\Omega s)}$
l	Length of field lines in m
μ	Absolute permeability in $\frac{Vs}{Am}$
A	Cross-section of the material in m^2

5.2.7 Force exerted on a conductor in the magnetic field – Lorentz force

A current-carrying conductor is deflected in the magnetic field. The direction of this deflecting force – referred to as Lorentz force – depends on the direction of the magnetic field (pole field) and on the current direction (conductor field) (see Figure 5.10).

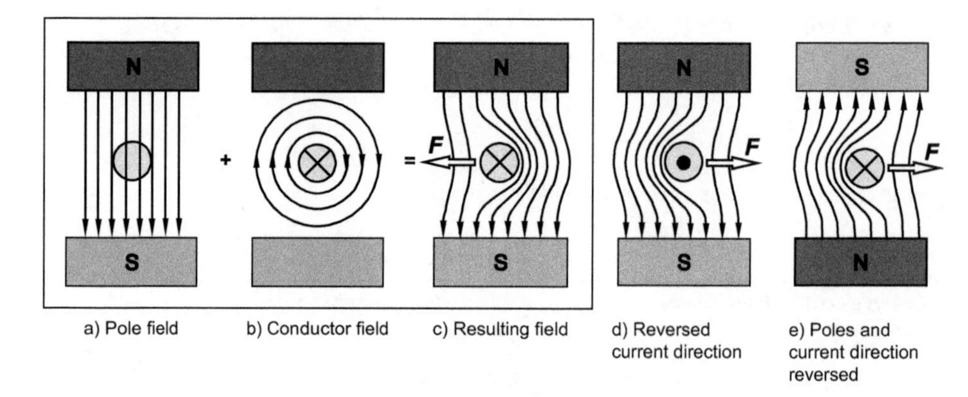

Figure 5.10: Current-carrying conductor in the magnetic field.

5.2.8 Left-hand rule for motors

The direction, in which a current-carrying conductor is deflected in a magnetic field, can be determined using the left-hand rule for motors (see Figure 5.12).

56 To mark the difference to current, the unit is sometimes labelled with "At" (ampere-turns).

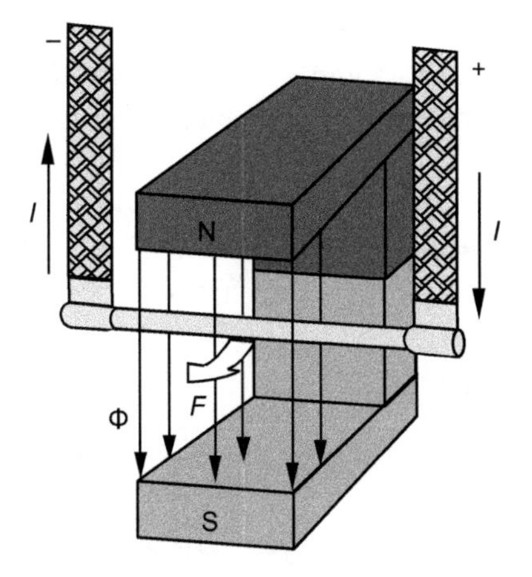

Figure 5.11: Lorentz force of a current-carrying conductor.

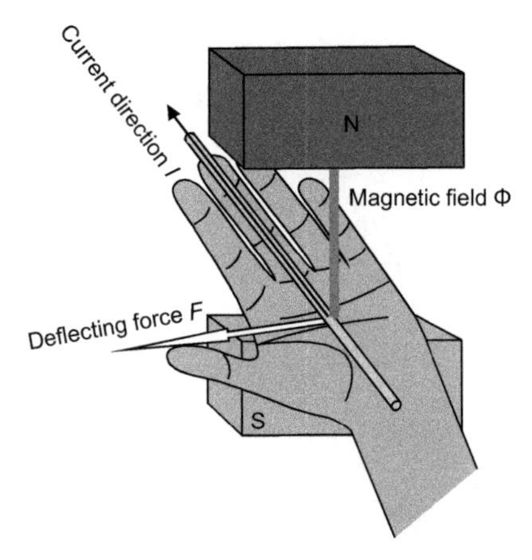

Figure 5.12: Left-hand rule for motors.

If the fingers of the flat hand point in current direction and the field lines, coming from the direction of the north pole, hit the palm of the hand, then the splayed thumb points in the direction of the deflecting force (Lorentz force).

Experiments have taught us that the Lorentz force F increases if

- the conductor current increases,
- the magnetic flux density increases,
- the effective conductor length extends, and
- that these correlations are linear.

Or to put it mathematically:

Lorentz force

$$\vec{F} = \vec{I} \cdot \left(\vec{B} \times \vec{l}\right) \cdot z$$

$\vec{F}$ Lorentz force in N
$\vec{I}$ Current in A
$\vec{B}$ Magnetic flux density in T (Tesla)
$\vec{l}$ Effective conductor length in m
z Winding number

The longer the effective conductor length (represented in the formula with $\vec{l} \cdot z$), the higher is the Lorentz force. This leads to the idea to use one conductor several times by turning it into a coil.

A current-carrying coil creates a magnetic north and south pole (electromagnet). In the magnetic field, the Lorentz force $\vec{F}$ (Figure 5.13) or a torque[57] acts on the

57 *Torque = Lorentz force · lever arm vector.*

electromagnet as long as the poles of the coil are directed towards the unlike poles of the magnetic field. If the coil was rotatable mounted and we reverse the current direction within the coil[58] in its present position, the coil would continue to spin in order to adjust itself towards the unlike poles of the magnetic field yet again. This is the operating principle of rotation in a DC motor. In such a motor, the alternation of current direction is performed by a commutator. The simplest version consists of two semi-cylinders on a shaft that are electrically insulated against each other and are contacted with brushes made of graphite (see Figure 5.14).

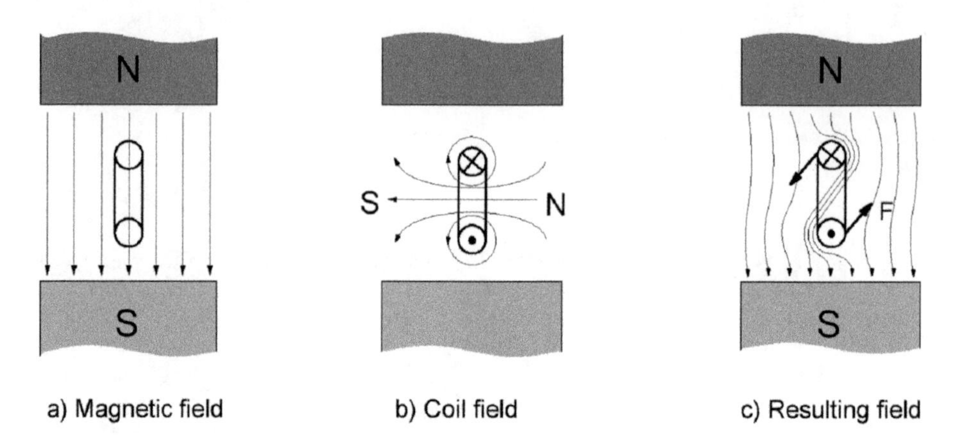

Figure 5.13: Current-carrying coil in the magnetic field.

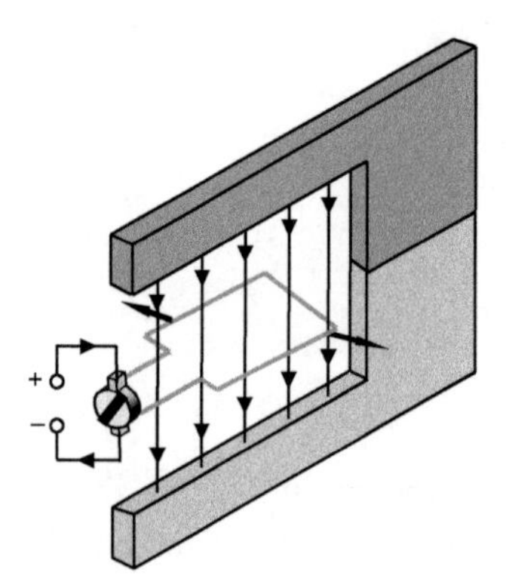

Figure 5.14: Conductor loop with commutator in the magnetic field.

58 In a direct current motor this happens in a commutator.

5.2.9 Induction

If a conductor or a coil is penetrated by an alternating magnetic flux, charges are shifted within the conductor or coil. In this way, a voltage called induction voltage is generated. There are two possibilities to create the alternating flux that leads to the generation of induction voltage in the conductor:
- by movement of the conductor[59] (dynamic induction = generator principle) or
- by alternating the flux in a motionless conductor (static induction = transformer principle).

5.2.10 Dynamic induction (generator principle)

When the magnetic flux within the loop changes due to the movement of a conductor loop in a magnetic field, a voltage is induced.

Experiments show that
- the direction of the induced voltage is proportional to the **direction of movement** and proportional to the **direction of the magnetic field** and that
- the level of voltage is proportional to **the speed of movement** of the conductor and
- proportional to the **number of turns** of the conductor loop.

To put it mathematically, the law relating to dynamic induction is formulated as follows:

Dynamic induction		
$v_i = \left(\vec{l} \times \vec{B}\right) \cdot \vec{v} \cdot z$	v_i	Induced voltage in V
	$\vec{B}$	Magnetic flux density in T (Tesla)
	$\vec{l}$	Effective conductor length in m
	$\vec{v}$	Conductor velocity in $\frac{m}{s}$
	z	Number of conductors / winding number

The direction of the current depends on the direction of movement of the conductor and on the direction of the magnetic field. It can be determined using the right-hand rule for generators:

59 When fixing the conductor and moving the magnetic field, a voltage is also induced. Therefore, it only depends on a relative movement of the conductor and the magnetic flux towards each other.

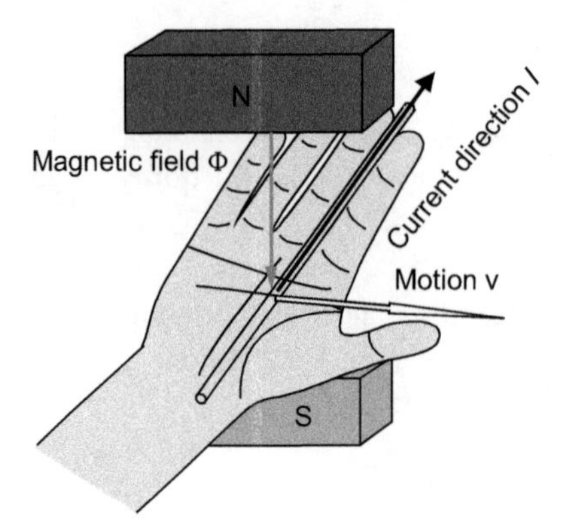

Right-hand rule for generators

If you hold the right hand in a way that the fieldlines meet the palm from the north pole and the splayed thumb points into the direction of movement, the induction current flows in the direction of the outstretched fingers.

Figure 5.15: Right-hand rule for generators.

The direction of the current caused by the induction voltage can also be determined using Lenz's law:[60]

Lenz's law
The current generated by an induction voltage is always directed against the cause of induction.

5.2.11 Static induction (transformer principle)

Experiments show that
- a voltage is induced in a coil if the penetrating magnetic flux changes,
- the voltage depends on the alteration speed of the flux change and that
- the induced voltage is proportional to the winding number.

To put it mathematically, the induced voltage v_i is formulated as follows:

Induced voltage	v_i	Induced voltage in V
$v_i = -N \cdot \frac{d\Phi}{dt}$	$\frac{d\Phi}{dt}$	Change in direction of the magnetic flux in $\frac{Wb}{s}$
	N	Winding number of the coil

In a test layout as illustrated in Figure 5.16, a voltage curve for v2 is the result of switching the current i1 on and off, as shown in Figure 5.17.

60 Heinrich F.E. Lenz, Russian physicist.

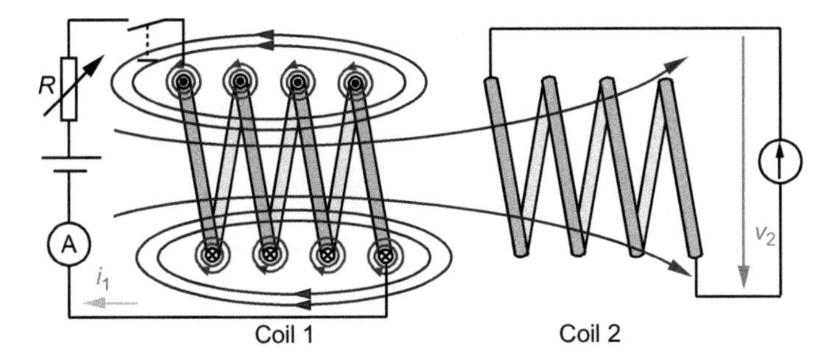

Figure 5.16: Static induction (transformer principle).

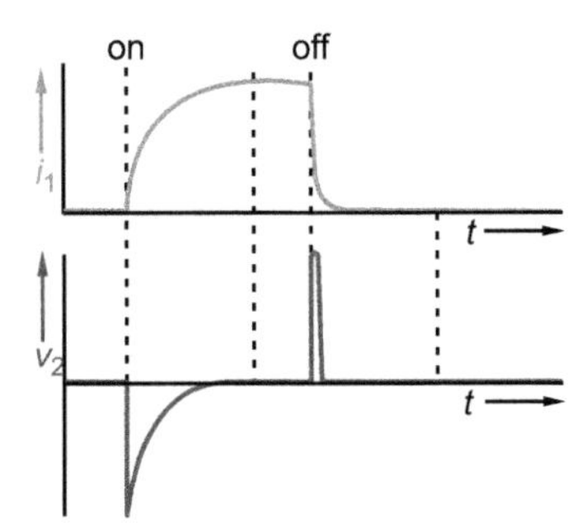

Figure 5.17: Voltage in coil 2 when turning the current in coil 1 on and off.

5.2.12 Self-induction

Induction not only happens in conductors penetrated by an (alternating) magnetic field of another conductor, but also in the conductor or the coil itself that generates said magnetic field. We can picture self-induction like this: Magnetic field lines "do not know" if the free electrons in the metallic conductor influenced by them are located in their own or in another conductor.

This has the following impacts on coils:

- If the coil is connected to an alternating voltage V, a self-induced voltage occurs in the coil, which counteracts the applied voltage (Lenz's law) thus decreasing the current consumption I. Outwardly the coil thus has a higher "resistance" – called inductive reactance.
- In case the coil is connected to a direct voltage V, the current in the coil does not immediately reach its full value. The increase of current and consequently the generation of the magnetic field are delayed. After switching off the current, the magnetic field of the coil dissipates. Again, this leads to a self-induction voltage which, however, is polarised in a way that the coil current continues to flow in the same direction and then slowly reduces to zero (Lenz's law).

For the inductance L and the induced voltage v_i the following correlations apply:

Inductance L	**Self-induced voltage v_i**
$L = N^2 \cdot \mu \cdot \frac{A}{l_m} = N^2 \cdot A_L$	$v_i = -L \cdot \frac{dI}{dt}$
L Inductance in H N Winding number of the coil μ Permeability in H/m or V*s/A*m l_m Average field-line length of iron in m A Area penetrated by the magnetic flux in m^2 A_L Coil constant, e.g. in mH according to manufacturer specifications	v_i Self-induced voltage in V L Inductance in H $\frac{dI}{dt}$ Current change according to time

5.2.13 Energy content W_{magn} of the magnetic field

The energy of the magnetic field of an inductance L (e.g. a coil) that is passed through by the current I is:

$$W_{magn} = \frac{1}{2} \cdot L \cdot I^2$$

Analogy:[61] Energy in the electric field $W_{el} = \frac{1}{2} \cdot C \cdot V^2$

5.3 Review questions

1) What are magnetic field lines and what is their direction?
   a) Sketch the resulting field line path of two parallel concordant current-carrying conductors. State the direction of the current and the field lines.
2) What does the relative permeability state?
   a) What is the difference between ferromagnetic, paramagnetic and diamagnetic materials?
   b) What is the ratio by which iron conducts the magnetic field better than air?
3) Explain the following terms:
   a) Remanence
   b) Coercivity
   c) Hysteresis loop
   d) Initial curve
4) What leads to the creation of Lorentz force?
   a) How can the direction of the Lorentz force be determined (detailed explanation)?
   b) How does the rotation of the rotor in a motor occur?

61 see page 80.

5) What is induction and what types of induction are there?
6) What is the energy content of the magnetic field?

5.4 Exercises

EXERCISE 5.1
A coil with a cross-section of the pole of 50 mm x 30 mm should induce an induction B of 0.8 T. Calculate the magnetic flux ϕ in Wb.

EXERCISE 5.2
On a nonmagnetic ring coil with an inner diameter of 50 mm and an outer diameter of 60 mm is a copper wire with 300 windings. Calculate the magnetic field strength H in A/m, the Induction B in T and the magnetic flux ϕ in Wb when a current of 1.5 A flows through the winding conductor.

EXERCISE 5.3
A magnet with a magnetic flux of 0.2 mWb is moved with constant speed into a coil with 60 windings within 3 ms. Calculate the induced voltage on the coil taps.

EXERCISE 5.4
A copper bar is moved with a speed v of 20 m/s through a b = 10 cm wide magnetic field. The amperemeter with an inner resistance R_i of 12 Ohm indicates a current I of 0.5mA. Calculate the Induction B of the magnetic field.

EXERCISE 5.5
A conductor is placed within a 5 cm wide horseshoe magnet with an Induction of 0.08 5T. When a current of 30 A flows through this conductor how big is the induced force F in N on the conductor?

6 Electrochemistry

6.1 Basic electrochemical concepts

With special regard to electrical engineering, this chapter covers the branch of electrochemistry that deals with the generation and storage of electric current. The electrochemical oxidation and reduction reactions take place at the phase boundaries of the electrode and the electrolyte.

Galvanic cell

Chemical energy is transformed into electrical energy, current is produced, and electrochemical reactions take place spontaneously (negative free enthalpy).

Galvanic cells are categorised into three subgroups:

- Primary cells
- Secondary cells
- Fuel cells

Electrolytic cell

Electric energy is transformed into chemical energy. Two electrodes made of electron-conducting material, and the electrolytes with ion conductivity are conductively connected[62] to each other. At the two spatially separated electrodes electrochemical reactions take place.

Half-cell

A half-cell consists of one single electrode and an electrolyte into which the electrode is submerged (e.g. copper in a copper sulphate solution). If a (metal) electrode is submerged into a metal salt solution (same metal), the surface of the electrode becomes charged. With base metals (e.g. zinc) some metal atoms enter the solution and the released electrons stay on the surface of the electrode, which is now negatively charged. The positively charged metal ions remain bound to the negatively charged metal surface. Thereby an electrical double layer is formed where the negative and the positive charges balance each other out. When two half-cells are combined, a galvanic cell (connected through ionic conductor and electron conductor) is formed.

Anode: Electrode where the oxidation processes take place. In the galvanic cell, the anode is negative; in the electrolytic cell, the anode is positive.

Cathode: Electrode where the reduction processes take place.

62 Conductive signifies that the outer conductor circuit must be closed, and the electrochemical reactions on the resulting boundary layers between ionic conductor and electron conductor must be possible (e.g. through the presence of suitable basic materials).

https://doi.org/10.1515/9783110521115-006

In the galvanic cell, the cathode is positive; in the electrolytic cell, the cathode is negative.

For better understanding, Figure 6.1 shows the four electrochemical reaction types that are explained further in the course of this chapter.

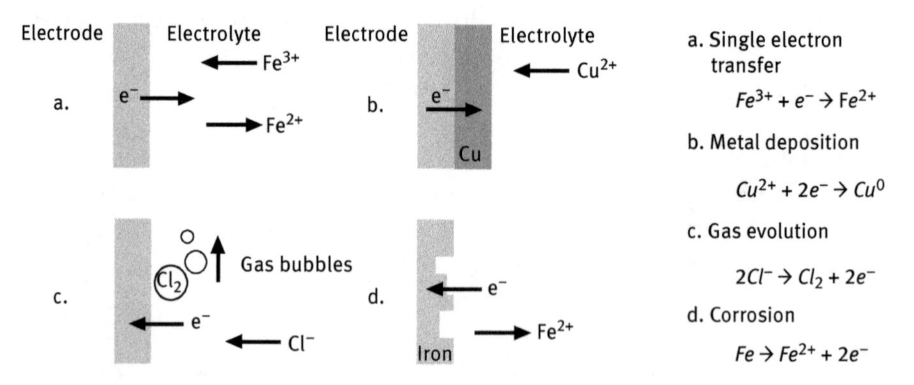

Figure 6.1: Four different electrochemical reaction types.

Electrolytes are **ionic conductors** and also consist of ions themselves.

They can be grouped into:
- acids; e.g. HCl (hydrochloric acid), H_2SO_4 (sulphuric acid)
- bases; e.g. KOH (potassium hydroxide), NaOH (sodium hydroxide)
- salts; e.g. NaCl (sodium chloride; salt), $CuSO_4$ (copper sulphate)

During **dissociation** the ionic lattice is split up into positive and negative ions. Only **dissociated electrolytes** are ionic conductors. Hence, especially aqueous solutions and molten salts are used. We differentiate between strong and weak electrolytes. Strong electrolytes are dissociated completely, whereas weak electrolytes are only dissociated very little. The electrons migrate through an external load circuit from the negative electrode to the positive one (electron flow). Inside the cell, the charged ions migrate to the respective oppositely charged electrode through an electrolyte with ion conduction. A current flow is generated (ion current).

6.2 Electrolysis

Electrolysis is a redox reaction provoked by electric current (inversion of processes of a galvanic cell); reduction and oxidation happen spatially separated. The electric current (direction of movement of positive charge carriers) is fed in through the **anode**[63] (positive pole) and is discharged at the **cathode** (negative pole). Between the two electrodes an electric field and ionic migration occurs. The negatively charged ions (**anions**) migrate to the anode where they are **oxidised.** During this process, electrons are emitted. The positive ions (**cations**) migrate towards the cathode where they are **reduced.** During this process, electrons are absorbed. Electrolysis is used to produce hydrogen and extract (pure) metals, among others.

6.2.1 Electrolysis of water

During the electrolysis of water, hydrogen (H_2) is produced at the cathode and oxygen (O_2) is produced at the anode. Through the addition of e.g. caustic potash solution, sulphuric acid or sodium chloride water becomes an electrolyte with ion conductivity. H_3O^+ and OH^- ions are created (also through autoprotolysis). The positive ions (H_3O^+, cations) move towards the cathode (negative) where they absorb an electron, producing H_2. The negative ions (OH^-, anions) move towards the anode (positive) where they emit an electron, producing O_2. The total electric charges of the protons and electrons amount to zero (in the closed system).

The complete partial reactions are as follows:

$$\text{Autoprotolysis:}\quad 2H_2O \rightarrow H_3O + OH^-$$

$$\text{Anode:}\quad 2OH^- \rightarrow \frac{1}{2}O_2 + H_2O + 2e^- \ \text{ or } \ 3H_2O \rightarrow \frac{1}{2}O_2 + 2H_3O^+ + 2e^-$$

$$\text{Cathode:}\quad 2H_2O + 2e^- \rightarrow H_2 + 2OH^- \ \text{ or } \ H_3O^+ + e^- \rightarrow \frac{1}{2}H_2 + H_2O$$

$$\text{Overall reaction:}\quad H_2O \rightarrow H_2 + \frac{1}{2}O_2$$

In the fuel cell (e.g. PEFC), discussed later in this chapter, the reactions take place in reverse order.

6.2.2 Extraction of (pure) metals

To produce (pure) metals (e.g. platinum, copper), the “contaminated” metal is used as anode and disintegrates during electrolysis. The pure metal deposits itself at the cathode. The cleaning of copper with sulphuric $CuSO_4$ solution as electrolyte serves as an example for this.

63 The designation of the electrodes in a cell changes with the direction of the current flow.

$$\text{Anode (raw copper): } Cu \rightarrow Cu^{2+} + 2e^-$$

$$\text{(contamination): } Fe \rightarrow Fe^{2+} + 2e^-$$

$$\text{Cathode (high-grade copper): } Cu^{2+} + 2e^- \rightarrow Cu$$

Iron is oxidised and enters the solution as cation; simultaneously copper is deposited at the cathode. If the raw copper also contains nobler metals than copper (e.g. gold), the former are not oxidised and sink to the bottom as anode slime.

6.2.3 Electrochemical corrosion

The electrochemical corrosion occurs when two different metals clash with a liquid with electrolytic conductivity (or stray current). Thereby, the less noble metal is decomposed, as there is a voltage difference and consequently a current flow (ionic current in the electrolyte). To prevent corrosion, a protective coating (e.g. Cr) made of a less noble metal than the relevant metal can be applied. Several metals automatically form an oxide layer to protect the metal below (e.g. Cu, Pb, Al).

6.3 Faraday's law

M. Faraday (English physicist 1791 – 1867) describes the amount of substance deposited in an electrolyte bath as follows:

$$m = ECE \cdot I \cdot t$$

$$Q = I \cdot t$$

$$ECE = \frac{M}{z \cdot F}$$

m Amount of substance in g
ECE Electrochemical equivalent in g/As (Material constant)
I Current in A
t Time in s
Q Charge in A·s
M Molar mass in g/mol, (gold: M = 196.97 g/mol)
z Valency (e^- transferred per ion), oxidation state (e.g. gold z = 1)
F Faraday constant F = 96485 As/mol

To calculate how long it takes to electrolytically deposit a certain mass of a substance at constant current, the following formula is used:

$$t = \frac{m \cdot z \cdot F}{M \cdot I}$$

Table 6.1: Electrochemical equivalents of certain metals.

Substance	Equivalents in g/As
Silver	$1.118 \cdot 10^{-3}$
Copper (II)	$0.329 \cdot 10^{-3}$
Nickel (II)	$0.304 \cdot 10^{-3}$
Zinc	$0.339 \cdot 10^{-3}$
Chromium (III)	$0.180 \cdot 10^{-3}$

6.4 The electrochemical series

Each substance that is in contact with a solvent has a **solution pressure**; metals in particular, as they strive to enter the solution as ion. This happens until the pressure of the ions exerted on the metal through their charge reaches its solution pressure. The electrolyte has an **osmotic pressure** that counteracts the solution pressure.

Examples: With a **zinc electrode** in zinc sulphate, the solution pressure of zinc is greater than its osmotic pressure. Zinc therefore goes into solution as $Zn^{2+}$and charges itself **negatively.** A potential difference between metal and solution occurs.

With a **copper electrode** that is submerged into copper sulphate, the solution pressure of the copper is smaller than its osmotic pressure. This means that the copper is prevented from releasing ions into solution. Additionally, some cations of the electrolyte emit their charge to the surface of the electrode. The cations are neutralised (the solution as a whole becomes negative) and charge the metal **positively.** This process continues until there is a balance between solution pressure (metal emits ever more ions) and osmotic pressure. Again, this results in a potential difference [3].

The **polarities** and therefore also the potentials of different metals depend on the electrolyte used as well as its concentration and temperature. The potential rating of the different metals in electrolytes (one molar solution of the examined metal) are listed in the **electrochemical series.**

The potential between electrolyte and metal can only be measured as compared to a reference electrode. The potential of the reference electrode is presumed as zero. The electrochemical series is measured against a **standard hydrogen electrode**

(SHE). **Oxides** can also be found in the electrochemical series. Metal oxides are always positive towards their respective metals, because they release further electrons only with difficulty. Materials with a **negative voltage rating** towards the SHE are called **base** ("like" to oxidise), they already dissolve in diluted electrolytes. Materials with a **positive voltage rating** towards the SHE are called **noble** (do "not like" to oxidise), they dissolve less easily. The greater the distance of two metals on the electrochemical series, the higher the voltage between them.

$$\Delta E = E(cathode) - E(anode)$$

If we take the Daniell cell (copper and zinc electrode) as an example for this, we get a cell voltage of 1.1 V (0.35 V + 0.76 V). Thus you can determine the electromotive force (emf, cell voltage) with the **difference** between two numeric values.

Mnemonic for selected elements: (K – Na – Mg – Al – Zn – Sn – Pb – H – Cu – Ag – Au)

Katty's Naughty** cat **mingled** with **Alice**, **Zarina** and **Selina Plundering Her Cupboard** of **Silver** and **Gold

Nernst equation

The Nernst equation describes how the potential of a redox pair is influenced by the change in concentration of the reaction partner. This allows us to calculate the electromotive force of a galvanic cell.

$$E = E^0 + \frac{R \cdot T}{n \cdot F} \cdot ln\left(\frac{a_{ox}}{a_{red}}\right)$$

E Electrode potential in V
E^0 Standard electrode potential in V
R Gas constant; = 8.314 $\frac{J}{K \cdot mol}$ or 0.0821 $\frac{l \cdot atm}{K \cdot mol}$
T Temperature in K
n Charge number or Number of transmitted electrons
F Faraday constant; = 96485 As/mol or C/mol (= $\frac{J}{V \cdot mol}$)
a Activity (ox. = oxidised form; red. = reduced form)

6.5 Primary cells

! Primary cells provide electric energy through electrochemical transformation processes. They are **not rechargeable**, because the reactions taking place are irreversible. The electrolyte solution can be used up or the base metal can be decomposed/disintegrated. Various combinations of electrodes from the electrochemical series are selected to form a **galvanic cell** with an electrolyte. Thereby, the cell voltage depends on the voltage difference between the two materials.

Table 6.2: Electrochemical series at 25 °C; 101.3 kPa; pH=0; ion activities=1.

Element	oxidised form	+ z e^-	⇌ reduced form	$E°$	Element	oxidised form	+ z e^-	⇌ reduced form	$E°$
Fluorine (F)	F_2	+ 2 e^-	⇌ 2 F^-	+2.87 V	Indium (In)	In^{3+}	+ 3 e^-	⇌ In	−0.34 V
Oxygen (O)	H_2O_2 + 2 H_3O^+	+ 2 e^-	⇌ 4 H_2O	+1.78 V	Iron (Fe)	Fe^{2+}	+ 2 e^-	⇌ Fe	−0.41 V
Gold (Au)	Au^{3+}	+ 3 e^-	⇌ Au	+1.50 V	Sulphur (S)	S	+ 2 e^-	⇌ S^{2-}	−0.48 V
Chromium (Cr)	Cr^{6+}	+ 3 e^-	⇌ Cr^{3+}	+1.33 V	Nickel (Ni)	NiO_2 + 2 H_2O	+ 2 e^-	⇌ $Ni(OH)_2$ + 2 OH^-	−0.49 V
Oxygen (O)	O_2 + 4 H^+	+ 4 e^-	⇌ 2 H_2O	+1.23 V	Zinc (Zn)	Zn^{2+}	+ 2 e^-	⇌ Zn	−0.76 V
Platinum (Pt)	Pt^{2+}	+ 2 e^-	⇌ Pt	+1.20 V	Water	2 H_2O	+ 2 e^-	⇌ H_2 + 2 OH^-	−0.83 V
Palladium (Pd)	Pd^{2+}	+ 2 e^-	⇌ Pd	+0.85 V	Chromium (Cr)	Cr^{2+}	+ 2 e^-	⇌ Cr	−0.91 V
Silver (Ag)	Ag^+	+ e^-	⇌ Ag	+0.80 V	Vanadium (V)	V^{2+}	+ 2 e^-	⇌ V	−1.17 V
Iron (Fe)	Fe^{3+}	+ e^-	⇌ Fe^{2+}	+0.77 V	Manganese (Mn)	Mn^{2+}	+ 2 e^-	⇌ Mn	−1.18 V
Copper (Cu)	Cu^+	+ e^-	⇌ Cu	+0.52 V	Titanium (Ti)	Ti^{3+}	+ 3 e^-	⇌ Ti	−1.21 V
Iron (Fe)	$[Fe(CN)_6]^{3-}$	+ e^-	⇌ $[Fe(CN)_6]^{4-}$	+0.36 V	Aluminium (Al)	Al^{3+}	+ 3 e^-	⇌ Al	−1.66 V
Copper (Cu)	Cu^{2+}	+ 2 e^-	⇌ Cu	+0.35 V	Cerium (Ce)	Ce^{3+}	+ 3 e^-	⇌ Ce	−2.483 V
Hydrogen (H_2)	**2 H^+**	**+ 2 e^-**	**⇌ H_2**	**0 V**	Calcium (Ca)	Ca^{2+}	+ 2 e^-	⇌ Ca	−2.87 V
Iron (Fe)	Fe^{3+}	+ 3 e^-	⇌ Fe	−0.04 V	Strontium (Sr)	Sr^{2+}	+ 2 e^-	⇌ Sr	−2.89 V
Lead (Pb)	Pb^{2+}	+ 2 e^-	⇌ Pb	−0.13 V	Potassium (K)	K^+	+ e^-	⇌ K	−2.92 V
Nickel (Ni)	Ni^{2+}	+ 2 e^-	⇌ Ni	−0.23 V	Lithium (Li)	Li^+	+ e^-	⇌ Li	−3.04 V

6.5.1 Leclanché cell

For the Leclanché cell, a coal/manganese dioxide (MnO_2) cathode, a zinc anode and an ammonium chloride (NH_4Cl) solution as electrolyte are used. With this battery, zinc is oxidised and manganese dioxide is reduced. During the oxidation electrons are released. They migrate through the outer circuit from anode to cathode, thereby releasing power. In parallel, OH^- ions migrate from the cathode to the anode in the electrolyte.

Figure 6.2 illustrates the structure of a Leclanché battery.

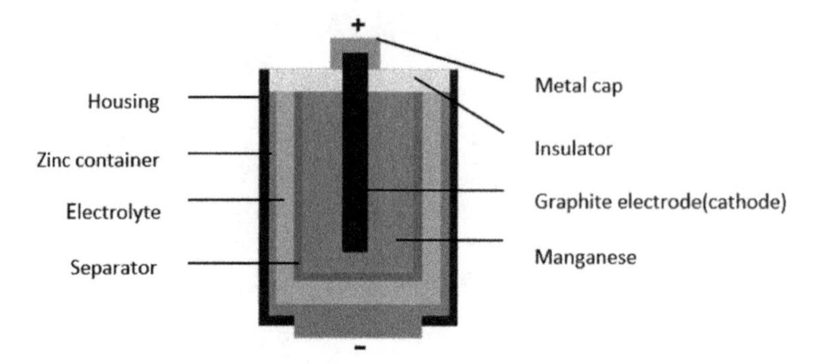

Figure 6.2: Structure of a Leclanché cell.

At the zinc electrode Zn^{2+} ions enter the solution and react with the ammonium chloride:

$$Zn \rightarrow Zn^{2+} + 2e^-$$

$$Zn^{2+} + 2NH_4Cl + 2OH^- \rightarrow [Zn(NH_3)_2]Cl_2 + 2H_2O$$

Manganese dioxide reacts with the electrolyte and the two electrons:

$$2MnO_2 + 2H_2O + 2e^- \rightarrow 2MnO(OH) + 2OH^-$$

Overall, the reaction is as follows:

$$Zn + 2MnO_2 + 2NH_4Cl \rightarrow [Zn(NH_3)_2]Cl_2 + 2MnO(OH)$$

During discharge the electrolyte solution is used up; the battery is not rechargeable.

The resulting cell voltage for the Leclanché cell is 1.5 V.

Further examples for primary cells:

- **Alkaline (manganese) cell**: Zinc powder as anode and manganese as cathode in caustic potash solution (1.5 V)

- **Silver-oxide (zinc) cell**: Zinc powder as anode and silver oxide as cathode in caustic potash solution, similar in structure to the alkaline battery (1.55 V)
- **Lithium manganese dioxide**: Lithium as anode, manganese dioxide as cathode, organic electrolyte, high voltages possible as lithium is very base (3 V)

6.6 Secondary cell

Secondary cells, or accumulators, are **rechargeable** multiple times. During the charging process, the accumulator acts as load, during the discharging process as generator. Electrochemical discharge reactions are generally reversible. Thus, multiple conversions from chemical to electric energy and vice versa can happen. During the discharging and charging process, oxidation and reduction processes alternately take place at the electrodes.

!

6.6.1 Lithium-ion battery

In particular in the last few years, the lithium-ion battery has gained ever more importance, as the use of lithium has considerable benefits compared to other elements.

As lithium has a very low equivalent weight and a high specific charge as well as a strongly negative standard electrode potential, lithium-ion batteries can reach high nominal voltages (3.6) and high energy densities (approx. 200–630 Wh/l).[64]

The operating principle of a lithium-ion battery is illustrated in Figure 6.3. The negative electrode generally consists of graphite into which lithium cations are intercalated during the charging process between the carbon layers, whereby the charge is absorbed by the carbon lattice.

Lithium intercalation reaction:

$$Li_xC_n \rightleftharpoons xLi^+ + xe^- + C_n \qquad \rightarrow \text{discharge} \leftarrow \text{charge}$$

The positive electrode generally consists of transition metal oxides in which lithium cations $LiMO_2$ (M=Co, Ni, Mn) are intercalated during discharge. Organic solvents (e.g. ethylene carbonate: diethyl carbonate = 1:1), serve as electrolytes. Conductive salts containing lithium (e.g. $LiPF_6$) are added to these electrolytes.

Overall cell reaction:

$$Li_{1-x}MO_2 + Li_xC_n \rightleftharpoons LiMO_2 + C_n \qquad \rightarrow \text{discharge} \leftarrow \text{charge}$$

64 Source: European Battery cell R&I Workshop – 11th January 2018.

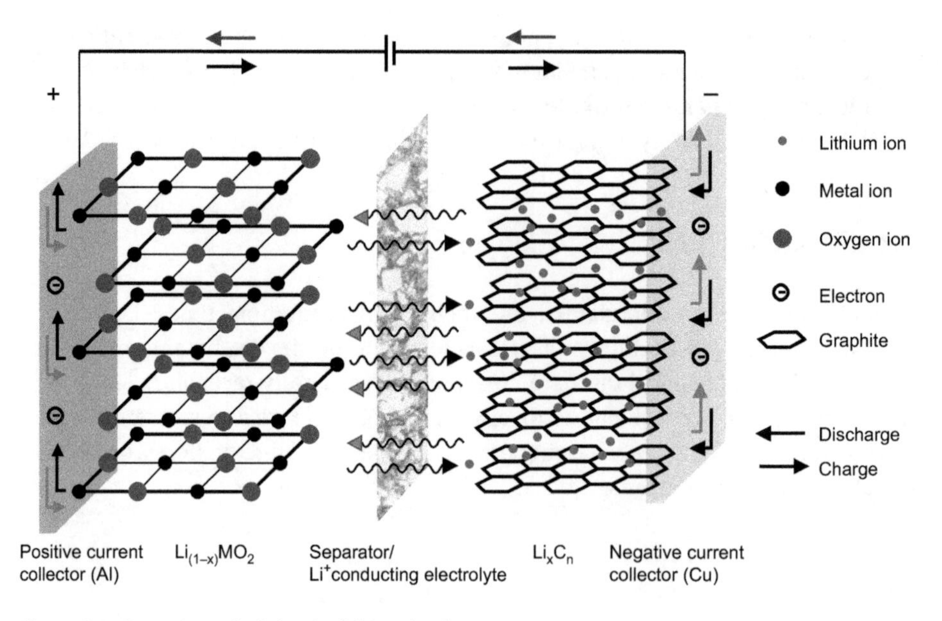

Figure 6.3: Operating principle of a lithium-ion battery.

Please note: In this type of accumulator, no oxidation or reduction reactions of Li/Li^+ happen, but instead the host lattice is reduced or oxidised. Li^+ ions are **only** intercalated or deintercalated but remain in cation form.

Further examples of secondary cells:

- **Lead accumulator**: Both electrodes consist of lead, coated with a layer of $PbSO_4$ or PbO_2 in a sulphuric-acid electrolyte; while charging, the positive electrode is transformed into PbO_2, the negative electrode consists of lead (2.04 V). During the discharging process, $PbSO_4$ forms at both electrodes.
- **Nickel iron, nickel cadmium and nickel metal hydride accumulators**: The positive electrode consists of NiO(OH) when charged; during discharge, it is transformed into $Ni(OH)_2$. Diluted potassium hydroxide solution is used as electrolyte (1.2 V); iron, cadmium or a hydrogen storage alloy function as negative electrode.

6.7 Fuel cell

! In a fuel cell the chemically bound energy of a continuously supplied fuel **is transformed directly into electrical energy** and heat. The most common form of the fuel cell – the hydrogen fuel cell – uses the energy set free in the usually explosive oxyhydrogen reaction (knallgas reaction). The reaction product is only water. The fuel cell itself is not an energy storage medium but a converter. However, considering the fuel cell with its hydrogen storage, this system as a whole can very well be counted as storage for electrical energy.

Hydrogen has a very high gravimetric energy density (specific energy): It contains **three times**[65] the energy of petrol. 1 kg of hydrogen equals 2.75 kg of petrol. The energy content of 1 Nm^3 (standard cubic meter) equals 0.34 l of petrol; 1 l of liquid hydrogen equals 0.27 l of petrol.

Fuel cells reach very high conversion efficiencies (approx. 60 %)[66] and are used in mobile, stationary and portable applications.

6.7.1 Structure and operating principle of a fuel cell

The fuel cell consists of an anode, a cathode and an electrolyte. At the anode, e.g. hydrogen oxidises. For each hydrogen atom, one electron is emitted. The resulting H^+ ions diffuse through the electrolyte from the anode towards the cathode. At the cathode, oxygen is reduced. Each oxygen atom formally gains two electrons and then reacts with two hydrogen protons to form a water molecule.

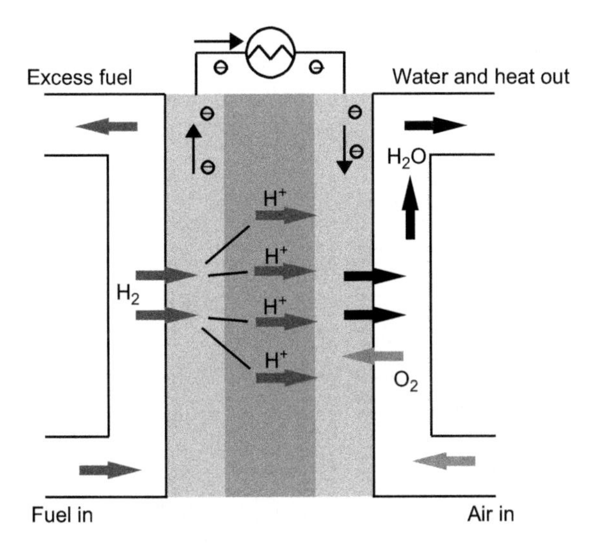

Figure 6.4: Operating principle of the hydrogen fuel cell.

65 Gravimetrically considered and with reference to the lower heating value (LHV).

66 The efficiency depends on the operating point.

6.7.2 General aspects of fuel cells

Ecological advantages

- No CO_x or NO_x pollution (if pure hydrogen is used, no greenhouse gases are emitted)
- The reaction product is pure water (with hydrogen as fuel)
- High efficiency
- No or only low noise pollution (due to additional aggregates)
- Hydrogen is potentially a clean "renewable" energy source.

Economic advantages

- Higher efficiency than thermal engines (e.g. ICE,[67] turbines, generators).
- Very high partial load efficiency
- Reaches high efficiency even in small units
- Direct generation of electrical energy
- No moving components, consequently a very long lifetime
- It is possible to use fuel cells in mobile, portable or stationary applications
- Hydrogen is stored more easily than electric current
- Hydrogen can be transported over long distances

Economic disadvantages

- Currently still a high price for fuel cells due to the use of expensive materials and the production on small scale
- For hydrogen fuel cells, a hydrogen infrastructure is required

6.7.3 Electrochemical process

We will explain the electrochemical process in a fuel cell that happens at the three-phase boundary (gaseous fuel – (moist) polymer membrane – catalyst/electrode) using the PEM (proton-exchange membrane) fuel cell as example:

Hydrogen is separated into protons and electrons at the platinum surface (anode). The resulting electrons are subsequently conducted through the electrode. Meanwhile, the protons move through the water stored in the membrane. The supply and discharge of reaction gases happen by means of porous carbon electrodes.

67 ICE = internal combustion engine.

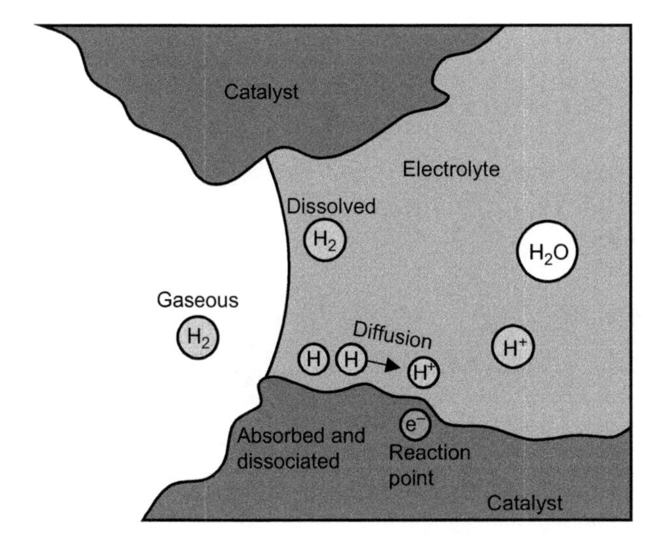

Figure 6.5: Representation of the chemical reaction at the three-phase boundary.

At the anode side, the fuel gas hydrogen is oxidised catalytically to $2H^+$:[68]

$$H_2 \rightarrow 2H^+ + 2e^-$$

The resulting electrons reach the cathode through the external load circuit where the oxidising agent oxygen is reduced catalytically together with the protons (H^+):

$$\frac{1}{2}O_2 + 2\,e^- + 2H^+ \rightarrow H_2O$$

Depending on the type of electrolyte, either the protons (H^+, H_3O^+) migrate from the anode to the cathode (PEFC, PAFC)[69] or the anions (OH^-, CO_3^{2-}or O^{2-}) migrate the other way round (AFC, MCFC, SOFC).[70]

The overall reaction

$$H_2 + \frac{1}{2}O_2 \rightarrow H_2O$$

generates a cell voltage of 1.23 V.

68 H^+ should be seen as simplification. As a rule, no free protons H^+ occur. In aqueous surroundings e.g. H_3O^+ (hydronium ions) are present.

69 PEFC = Polymer Electrolyte Fuel Cell (=PEMFC Proton Exchange Membrane Fuel Cell); PAFC = Phosphoric Acid Fuel Cell.

70 AFC = Alkaline Fuel Cell; MCFC = Molten Carbonate Fuel Cell; SOFC = Solid Oxide Fuel Cell.

6.7.4 Energy balance/efficiency η_{max}

Energy balance

In a fuel cell, the chemical energy stored in the fuel is converted into electrical energy and heat energy. The chemical bond energy of the fuel is emitted during the combustion as reaction heat. This value is called enthalpy of reaction ΔH or heating value under standard conditions.

$$2\,H_2 + O_2 \rightarrow 2H_2O \qquad \Delta H^0 = -285.8\ kJ/mol$$

Ideal electrical efficiency of a fuel cell

The ideal (electrical) efficiency of a fuel cell is:

$$\eta_{max} = \frac{maximum\ amount\ of\ electrical\ work}{reaction\ heat\ of\ the\ fuel} = \frac{\Delta G^0}{\Delta H^0}$$

ΔH^0 Calorific value/heating value under standard conditions
ΔG^0 Free enthalpy of reaction under standard conditions

Examples (at 25° C)
- Hydrogen (Higher Heating Value; HHV):[71] 83 %
- Hydrogen (Lower Heating Value; LHV):[72] 94 %

The reversible cell voltage of the fuel cell can be calculated based on correlation:

$$\Delta G = -n \cdot F \cdot E$$

ΔG^0 Free enthalpy of reaction
n Number of transferred electrons
F Faraday constant; = 96485 As/mol or C/mol ($= \frac{J}{V \cdot mol}$)
E Reversible cell voltage (also referred to as open circuit voltage or open cell voltage; OCV) in V

6.7.5 Loss mechanisms and overpotential

Overpotential (polarisation) refers to the deviation of electrode potential during current flow as opposed to the equilibrium potential of the electrode.

71 The higher heating value is the chemically bound energy contained in a substance that is set free during combustion and subsequent cooling (25° C) as well as during condensation (liquid water).

72 The lower heating value considers only energy set free during the combustion of a substance (product water in the gaseous phase).

The losses occurring in a fuel cell are mainly the result of three causes:

- Charge-transfer overpotential
- ohmic losses
- Diffusion overpotential.

The charge-transfer overpotential (inhibited charge transfer) is determined by the reaction of the electron transfer. This is caused by the limiting velocity of the exchange of charges at the electrode/electrolyte interface. This mainly occurs with low current densities.

The ohmic losses are caused by the internal resistance of the cell. The voltage at the terminals of the fuel cell is reduced by the voltage dropping proportional to the current on the internal resistor.

The diffusion overpotential (inhibited mass transport) happens because of the slow mass transport (diffusion) to the electrode (=depletion of reactants at the electrode). It increases with higher current density.

Even before proper load operation mixed potential losses (through fuel crossover – diffusion of reaction gases through the electrolyte) and losses through internal currents (through non-ideal membranes) occur. These are illustrated in Figure 6.6 with the deviating value E^0_{real}.

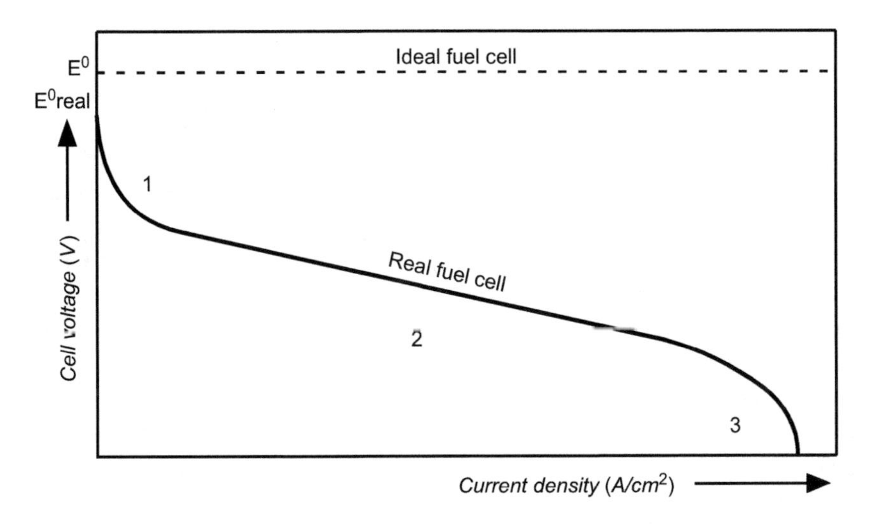

Figure 6.6: Current density-voltage characteristic of a fuel cell.

The operating behaviour of a fuel cell is described by means of the current (density)-voltage characteristic. Figure 6.6 illustrates the most important losses that occur.

E^0 is the maximum voltage that can theoretically be applied to a fuel cell in open circuit mode (open circuit voltage or OCV – open cell voltage). In reality, this voltage is decreased by losses.

Butler-Volmer equation

The exponential correlation of charge-transfer overpotential and current density is defined in the Butler-Volmer equation.

$$j = j_0 \left[e^{\frac{\alpha \cdot n \cdot F \cdot \eta_D}{R \cdot T}} - e^{-\frac{(1-\alpha) \cdot n \cdot F \cdot \eta_D}{R \cdot T}} \right]$$

j Current density in A/cm^2
j_0 Exchange current density in A/cm^2
α Symmetry factor $0 > \alpha > 1$; often 0.5) or Charge transfer coefficient
F Faraday constant = 96485 in C/mol or As/mol ($= \frac{J}{V \cdot mol}$)
η_D Charge-transfer overpotential in V
R Gas constant; = 8.31 in $\frac{J}{K \cdot mol}$; = 0.0821 in $\frac{l \cdot atm}{K \cdot mol}$
T Temperature in K

6.8 Electrochemical impedance spectroscopy

The electrochemical impedance spectroscopy (EIS) is used to determine material and system properties of electrochemical, electrical and electronic components. EIS determines the impedance, i.e. the alternating current resistance of the medium depending on the frequency of the measurement current flowing through. In a fuel cell EIS describes the ion conductivity of the electrolyte and the kinetics of the electrochemical reactions.

Two factors are important when measuring the impedance:

1. Time-invariant system: When measuring, the system must be constant. For example, when doing research on fuel cells, the set operating point (set current, set voltage) and the temperature have to remain the same.
2. Linearity: The characteristic curve of the system must be linear in the investigated area, otherwise there will be distortions. The measuring of a fuel cell, for example, has to be done in the linear range (ohmic range).

Figure 6.7 illustrates the operating principle of EIS by means of a fuel cell.

To record the impedance spectrum, first of all a stationary operating point (with batteries via discharge/charge, with fuel cells via generation of current/electrolysis) is set. Subsequently, a sinusoidal alternating current signal $\underline{I}$ with the

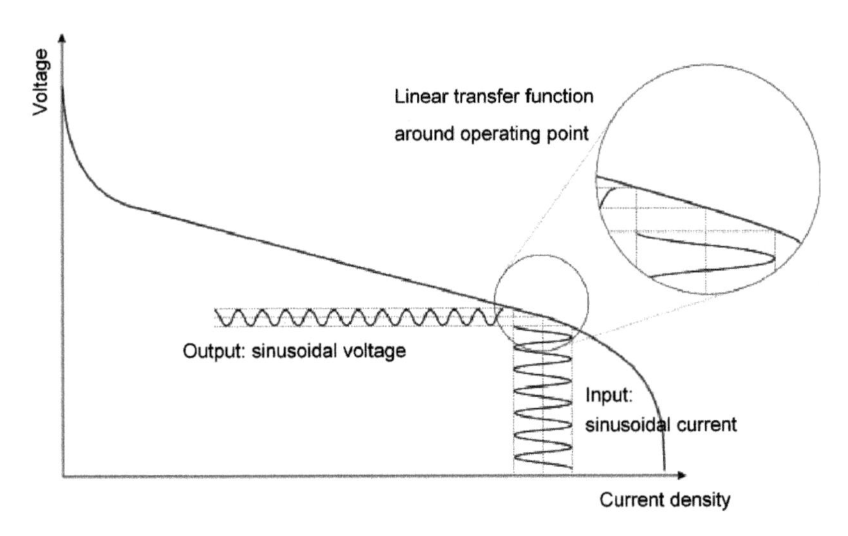

Figure 6.7: Operating principle of the electrochemical impedance spectroscopy.

known frequency f modulates the system. The current causes a voltage drop V (with phase angle ϕ) at the alternating current resistor $\underline{z}$ that is being measured.

Knowing $\underline{i}$ and the measured $\underline{v}$, the impedance $\underline{z}$ for the frequency f of the system can be calculated in accordance with

$$\underline{Z} = \frac{\underline{V}}{\underline{I}} = \frac{|\underline{V}| \cdot e^{j\varphi}}{|\underline{I}| \cdot e^{j0^\circ}} = |Z| \cdot e^{j\varphi} = |\underline{Z}| \cdot \cos\varphi + |\underline{Z}| \cdot \sin\varphi = Re(\underline{Z}) + j \cdot Im(\underline{Z})$$

with

$$|\underline{Z}| = \sqrt{Re(\underline{Z})^2 + Im(\underline{Z})^2} \varphi = arctan\left[\frac{Im(\underline{Z})}{Re(\underline{Z})}\right]$$

Now we know the impedance for a frequency value. To illustrate a spectrogram, the frequency f within a certain interval is changed and the respective calculated impedance values are recorded.

Representation of impedance spectra

The two most important representations for the EIS are the Nyquist and the Bode plot.

Nyquist plot

The real part of the impedance is applied at the abscissa, the imaginary part at the ordinate.

Advantage: The number and kinetics of the processes taking place can be represented.
Disadvantage: The dependence on the frequency is not reflected.

Bode plot
The frequency is applied to the abscissa logarithmically. The impedance $\underline{Z}$ and the phase angle φ are applied to the ordinate.

Table 6.3 shows the impedances with real and imaginary parts of the most important electrical components.

Table 6.3: Impedance with real and imaginary part of the most important electrical components.

Electrical component	$\underline{Z}$	Re($\underline{Z}$)	Im($\underline{Z}$)	φ
Resistor	R	R	0	0°
Capacitance (capacitor)	$\frac{1}{j\omega C} = -\frac{j}{\omega C}$	0	$-\frac{1}{\omega C}$	−90°
Inductance	$j\omega L$	0	ωL	90°

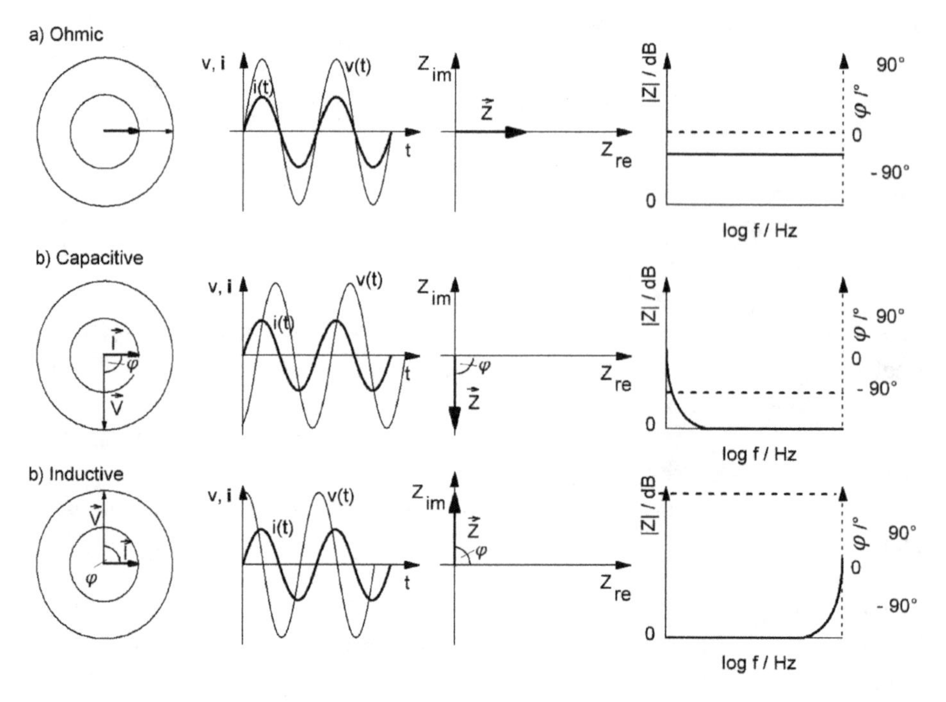

Figure 6.8: Illustration of current and voltage in a Nyquist and Bode plot for capacitor and inductance.

Figure 6.8 the vector diagrams as well as the corresponding Nyquist and Bode plots for basic components. To analyse impedance spectra, we need suitable electric equivalent schematics. They should be selected in a way that they describe physical and chemical processes.

6.8.1 Cyclic voltammogram

Cyclic voltammetry (CV) determines electrode processes, reaction products and reversibility of electrochemical reactions. During CV measurements a triangle voltage is applied at the **working electrode** (WE, e.g. platinum). The resulting current flow is recorded as function of the working electrode potential. The measurement of the electrode potential is performed by a **reference electrode** (RE, e.g. silver-silver chloride) with known potential.

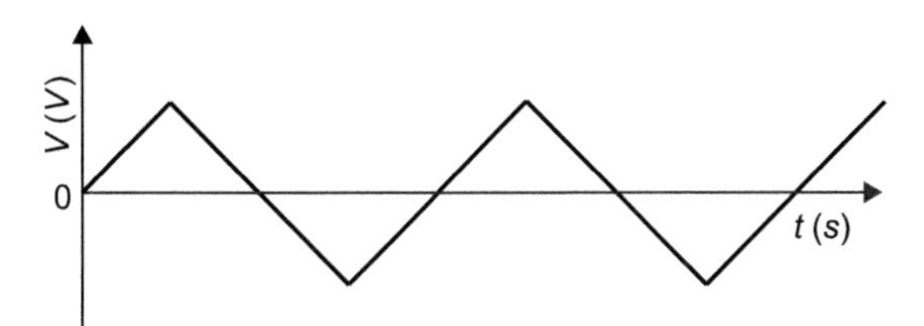

Figure 6.9: Triangle voltage.

As of a certain voltage, current flow occurs and the substance is oxidised at the electrode. If the (triangle) voltage decreases, fewer particles are oxidised. When the voltage reaches negative values, a reduction of the previously oxidised particles takes place.

For CV measurements with greater currents caused by higher voltage feed rates (speed with which the voltage is varied) or electrolytes with high resistances, a counter electrode (CE) as third electrode is used. The current is directed through the counter electrode (measurement of current flow between WE and CE); the reference electrode with high impedance is used for the voltage measurement between WE and RE. The working electrode and the counter electrode alternately assume the function of anode and cathode due to the voltage curve. We talk about a three-electrode setup.

In the case of e.g. an anodic current flow occurring at the working electrode due to oxidation, a reduction process happens at the counter electrode producing cathodic current.

$$\text{Anode:}\quad B_{red} \rightarrow B_{ox} + e^-$$

$$\text{Cathode:}\quad B_{ox} + e^- \rightarrow B_{red}$$

If the redox reaction is reversible, the shape of the cyclic voltammogram looks approximately like in Figure 6.10.

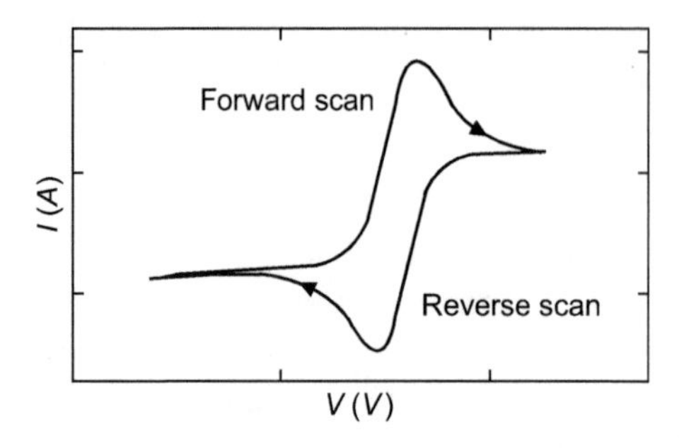

Figure 6.10: Schematic representation of a cyclic voltammogram of a reversible redox reaction.

During the forward scan (increasing triangle voltage) B_{red} is oxidised; during the reverse scan B_{ox} is reduced. Symmetrical current peaks occur at the reduction as well as the oxidation. Each particle produced in the forward scan is retransformed into the starting compound during the reverse scan.

Working electrode WE: Electrochemical processes take place there. In order to measure the potential, it is necessary to use a RE (e.g. standard hydrogen electrode, SHE).

Counter electrode CE: Current flows between CE and WE. Together with the WE (+electrolyte) the CE constitutes an electrochemical cell. If the anode is the WE, the CE is the cathode. The latter generally consists of the same metal as the working electrode.

Reference electrode RE: It is used as electrode (reference value of the potential) which enables a voltage measurement of the WE. Voltage is a potential difference and can only be measured with respect to a reference point (the potential of which is set to zero) – the RE represents this reference point.

6.9 Review questions

1) How is an electrical double layer created?
2) What are anode and cathode during an electrolysis? Which processes occur at the anode/cathode?
3) Describe the processes happening during water electrolysis and state the partial reactions.
4) What does it mean if a metal is referred to as noble or base metal? Give an example.

5) Leclanché cell
   a) Describe the structure of the primary cell.
   b) What reactions happen at the electrodes?
   c) What is the overall reaction?
6) Which primary cells do you know? Name 3 representatives and label the cathodes and anodes.
7) Lithium-ion battery
   a) Describe the structure of the lithium-ion battery.
   b) Which reactions happen during charge and discharge?
   c) What is the overall reaction?
8) Which secondary cells do you know? Name 3 representatives and state the electrode materials.
9) Cyclic voltammogram
   a) What is cyclic voltammetry used for?
   b) Briefly describe the operating principle.
   c) What is the three electrode setup?
   d) Explain the terms working electrode, counter electrode and reference electrode.
10) Fuel cell
   a) What type of fuel cells are you familiar with?
   b) Name the advantages and disadvantages of the use of fuel cells.
   c) Explain the processes at the electrodes (in a PEMFC).
   d) What is the three-phase boundary?
   e) What is the level of the ideal efficiency of the fuel cell?
   f) What losses can happen when using a fuel cell?

6.10 Exercises

Exercise 6.1

During an electrolysis, 6.3 g of copper is discharged at the cathode from a copper sulphate solution.

a. How long do we have to electrolyse the solution using a current of 0.1A until this amount is reached (the molar mass of copper is 63.55 g/mol)?
b. What level of current is needed if in one second 1.12 mg of silver are to be deposited from a silver nitrate solution (the molar mass of silver is 107.87 g/mol)?

Hint: $n = m/M$

Exercise 6.2

Two half-cells are combined to a galvanic cell (Daniell cell). The concentrations of the two half-cells are c (Zn/Zn^{2+}) = 0.35mol/l und $c(Cu/Cu^{2+})$ = 0.2mol/l.

The concentration can be considered as activity.
Hint:

$$E^0: \quad Cu^{2+} + 2e^- \rightleftharpoons Cu(0.35V) \qquad Zn^{2+} + 2e^- \rightleftharpoons Zn(-0.76V)$$

a. In which directions do the electrons flow?
b. Which voltage occurs?

EXERCISE 6.3
The exchange current density is $1.00 \cdot 10^{-5} \frac{A}{cm^2}$ for the reaction $H^+ + e^- \rightarrow \frac{1}{2}H_2$ in nickel at 25° C.
What level of current density is required to generate an overpotential of 0.100 V ($\alpha = 0.5$)?

7 Alternating current technology

Alternating current technology is an important area of electrical engineering as the public electricity supply happens through AC/three-phase AC technology. The high-power consumption in the public supply (which in Europe is a three-phase supply) is covered through power generation in power plants.[73] A generator creates sinusoidal alternating voltages. In order to transmit electric power over a great distance with little losses (power loss $P = I^2 \cdot R$), the current I must be small. The current is low at the same power ($P = V \cdot I$), if the voltage, in contrast to the current, is increased.[74] The level of voltages is technically adjusted with transformers in a simple and efficient way ("transformed"). Transformers work with alternating currents due to their mode of action.[75] Depending on the distance for the power to be bypassed and the power to be distributed, different specific voltage levels are chosen. By means of alternating current technology electrical energy can be produced and distributed easily. You can directly operate low-maintenance and technically simple engines (asynchronous motors) on a three-phase supply (three-phase current system). Before the current reaches the load, the voltages are transformed to a so-called low voltage level (400 V/230 V). Due to these advantages, AC technology has become established in the public electricity network. Direct current technology is, at the moment, mainly limited to electronics.[76]

Definition of terms

An alternating voltage constantly changes between a positive and a negative maximum value (peak amplitude $\hat{v}$ peak value or amplitude[77]). Similar applies for the current driven through the load by alternating voltage.

In correlation with alternating current technology further terms are required, all of which are represented in Figure 7.1.

73 or more precisely they transform mechanical kinetic energy into electrical energy.

74 Rule of thumb: 1 kV per km line length.

75 Static induction (a changing magnetic field forms an induction voltage in a static winding)

76 With the breakthrough of power electronics, energy transmission with direct current technology has recently been gaining in importance. In converter stations voltages up to several 100 kV and powers up to the GW range can be implemented. By means of high-voltage direct current (HVDC), current can be transmitted over long distances (up to 1,000 km) in a highly efficient way (no losses due to capacitive and inductive line coatings like in alternating current technology).

77 Deviating from that, in some publications or with some devices (e.g. functional generator), the peak-peak value is simply called "amplitude".

https://doi.org/10.1515/9783110521115-007

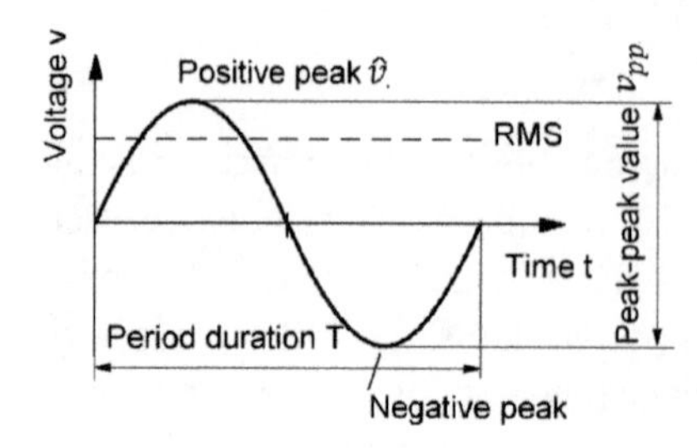

Figure 7.1: Terms in alternating current technology.

Advantages of alternating current compared to direct current:
- direct transformability
- simple energy transmission over long distances
- construction of effective machines (asynchronous motor, transformer) is possible
- well switchable

Conventional notation

Instantaneous values (time-dependent quantities) are represented with lower case letters. Effective values and time-independent quantities are represented with capital letters.
- Peak value (maximum) $\hat{v}$, $\hat{i}$, v_{max}, i_{max}
- Peak-peak value v_{pp}, i_{pp}
- Instantaneous value[78] v, i
- Rectified value V, I_m
- Effective value V, I

Fundamental differences to direct current technology

With alternating current technology, there are, additionally to the ohmic resistance ("active resistance"), also inductive and capacitive reactance that cause a **phase shift** between current and voltage. Therefore, the following applies: **ELI** the **ICE** man
- Inductance: Current lags the voltage by 90° – With an inductor (L), the emf (E) is ahead of the current (I).
- Capacitor: Voltage lags the current by 90° – With a capacitor (C): the voltage emf (E) is behind the current (I)

Please note: We have to differentiate the terms inductance and capacitance from the real components coil and capacitor. The latter are lossy and show an ohmic part additionally to the reactance (which causes the phase shift to be smaller than the respective 90°).

78 An instantaneous value is a time-dependent quantity.

Current through an inductance is called “inductive reactive current”; the voltage is called “inductive reactive voltage”. Correspondingly, when capacitance is concerned, current and voltage are called “capacitive reactive current” or “capacitive reactive voltage”.

7.1 Vector diagram of sinusoidal quantities

The **instantaneous values** of sinusoidal voltages or currents can be described with

$$v(t) = \hat{v} \cdot \sin\left(\underbrace{\omega \cdot t}_{\hat{=}a} + \varphi\right)$$

$$i(t) = \hat{i} \cdot \sin\left(\underbrace{\omega \cdot t}_{\hat{=}a} + \varphi\right)$$

whereby ω is the angular frequency ($\omega = 2 \cdot \pi \cdot f$) and φ is the zero-phase angle (=phase position).

The corresponding time diagrams $v(t)$ or $i(t)$ give a clear overview about the temporal sequence of the voltages and currents. The voltages and currents can thus be added, subtracted, as well as multiplied and divided. Therefore, in principle, all values of mixed alternating current circuits can be calculated. However, the analytic solution of trigonometric connections becomes time-consuming and confusing. In practice, the solution or representation and calculation with so-called vectors is preferred.

Sinusoidal processes can be symbolically represented with a (in a mathematically positive sense[79]) rotating vector (see Figure 7.2). The length of the vector corresponds to the peak value of the alternating quantity ($\hat{v}, \hat{i}$); the rotation frequency of the vector corresponds to the angular frequency ω of the alternating quantity. Temporal shifts ($\hat{=}$ phase shifts) from a mostly arbitrarily chosen temporal zero point are indicated with a zero-phase angle. The zero-phase angle can be positive or negative. The phase shift between two alternating quantities equals the difference of their zero-phase angle; if they are positive or negative depends on the chosen reference value.

79 Anticlockwise.

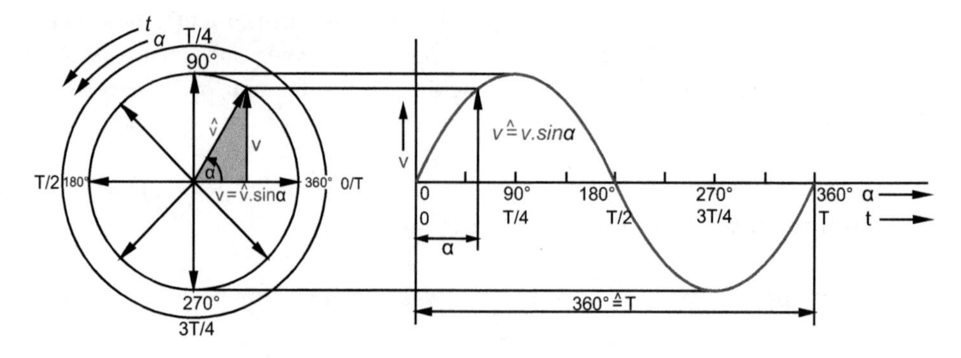

Figure 7.2: Vector diagram and curve diagram of a sinusoidal voltage.

Fixed vector

It is assumed that in a system with several vectors, all frequencies and therefore angular frequencies are the same (in practice this is usually the case[80]), the vectors rotate with the same rotation frequency, but the phase shift between the vector (and alternating quantities) stays the same. For the abovementioned reason and because in practice the effective value of an alternating quantity is most important, we can use non-rotating fixed vectors. The length of the vector corresponds to the effective value of the alternating quantity; the phase position is represented by the zero phase angle. With fixed vectors, all possible calculation operations for vectors can be performed graphically as well as mathematically.

7.2 Characteristic quantities in alternating current technology

7.2.1 Frequency f, time period T

The frequency f is the number of periodic occurrences per unit of time. In alternating current technology frequency in particular is identified as number of periods per second. The duration of a full period is called time period T.

The frequency f is the reciprocal of the time period T:

$$f = \frac{1}{T}$$

f Frequency in $\frac{1}{s} = Hz$
T Time period in s

80 ... because the electric supply network has a given frequency (in Europe, Asia, Australia, Africa 50 Hz and in North and Central America and Japan 60 Hz) and, therefore, all involved voltages and currents have the same frequency.

7.2.2 Phase shift φ

In an alternating-current circuit with only ohmic resistors as loads, the voltage and current are in phase which means they reach their peak values or zero-crossings simultaneously.

If there are also inductive or capacitive resistors in an alternating-current circuit, voltage and current are generally not in phase (i.e. $\varphi_u \neq \varphi_i$) that means they do not reach their peak values at the same time.

For the phase shift φ the following applies:

$$\varphi = \varphi_u - \varphi_i$$

7.3 Effective value (RMS)

The effective value (root mean square, RMS[81]) of an alternating current is as high as a direct current with the same heating effect.

Calculating, measuring and working with effective values has the great advantage that one does not have to consider the sinusoidal instantaneous values. This leads to simplifications when using AC technology. The instantaneous values, in practice, are often not important as it does not matter e.g. with electric heating if the level of current, that causes the heating effect, alternates sinusoidally or if it is a direct current with the same heating effect (definition of the effective value!). For a load, the root mean square (=effective value) of power is significant.

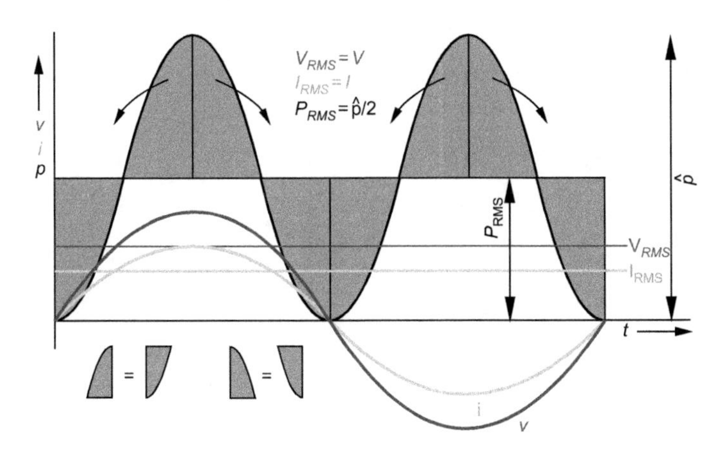

Figure 7.3: Effective value (RMS).

81 RMS is the abbreviation for **R**oot **M**ean **S**quare.

Mathematical derivation:

When comparing the power of an ohmic resistor passed through by direct current in one case and by alternating current in the other, the following considerations apply:

In the case of direct current, the following applies for the power:

$$P = V \cdot I = I^2 \cdot R = \frac{V^2}{R}$$

Correspondingly, in the case of alternating current, the following applies:

$$P(t) = i^2(t) \cdot R$$

If the sinusoidally alternating current $i(t) = \hat{i} \cdot \sin(\omega t)$ is squared, this results in:

$$i^2(t) = \hat{i}^2 \cdot \sin^2(\omega t)$$

The temporal sequences of the quantities are represented in Figure 7.4.

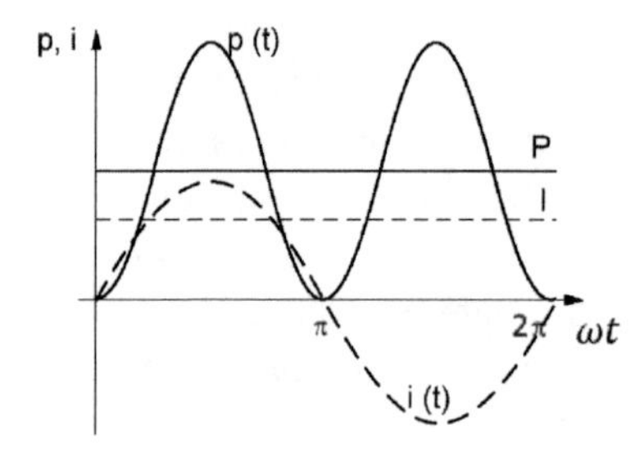

Figure 7.4: Effective value.

The mean value of the curve $i^2(t)$ is a direct quantity[82], it is designated as I^2. For the integral over the time period T, this results in:

$$I^2 = \frac{1}{T} \cdot \int_0^T i^2(t) \cdot dt = \frac{\hat{i}^2}{T} \cdot \int_0^T sin^2(\omega t) dt = \frac{\hat{i}^2}{\omega T} \int_0^T sin^2(\omega t) \cdot d(\omega t) = \frac{\hat{i}^2}{2}$$

For I this leads to:

$$I = \sqrt{\frac{1}{T} \cdot \int_0^T i^2(t) \cdot dt} = \sqrt{\frac{\hat{i}^2}{2}} = \frac{\hat{i}}{\sqrt{2}}$$

The resulting current value for I is called root mean square or **effective value.**

82 time-independent quantity.

Table 7.1: Effective value, arithmetic mean and crest factor[83] of various curve shapes.

Voltage	Effective value V	Arith. mean $\bar{v}$	Crest factor F_c	Voltage	Effective value V	Arith. mean $\bar{v}$	Crest factor F_c
sine-wave voltage	$\frac{\hat{v}}{\sqrt{2}}$	0	$\sqrt{2}$	sawtooth voltage	$\frac{\hat{v}}{\sqrt{3}}$	0	$\sqrt{3}$
rectangular alternating voltage	$\hat{v}$	0	1	triangular pulse	$\frac{\hat{v}}{\sqrt{3}}$	$\frac{\hat{v}}{2}$	$\sqrt{3}$
square pulse ($t_i = t_p$)	$\frac{\hat{v}}{\sqrt{2}}$	$\frac{\hat{v}}{2}$	$\sqrt{2}$	sawtooth pulse	$\sqrt{\frac{t_i}{3 \cdot T}} \cdot \hat{v}$	$\frac{t_i \cdot \hat{v}}{2 \cdot T}$	$\sqrt{\frac{3 \cdot T}{t_i}}$
square pulse 2	$\sqrt{\frac{t_i}{T}} \cdot \hat{v}$	$\frac{t_i \cdot \hat{v}}{T}$	$\sqrt{\frac{T}{T_i}}$	pulsating DC voltage	$\frac{\hat{v}}{2}$	$\frac{\hat{v}}{\pi}$	2
triangle voltage	$\frac{\hat{v}}{\sqrt{3}}$	0	$\sqrt{3}$	pulsating DC voltage 2	$\frac{\hat{v}}{\sqrt{2}}$	$\frac{2 \cdot \hat{v}}{\pi}$	$\sqrt{2}$

t_i pulse duration t_p pause duration T period duration

Considering $P = \frac{V^2}{R}$, an analogue derivation can be performed for voltage. We receive:

$$\boldsymbol{V = \frac{\hat{v}}{\sqrt{2}}}$$

If there is no additional indication regarding the specification of alternating voltage or alternating current, it is always the effective value that is intended. The alternating voltage of 230 V commonly used in households is also designated by its effective value. According to the nomenclature, effective values are easily recognised by the capital letters used as symbols.

83 The crest factor is given by: $F_c = \frac{\hat{v}}{V}$ while $\hat{v}$ refers to the peak value and V to the effective value of the voltage.

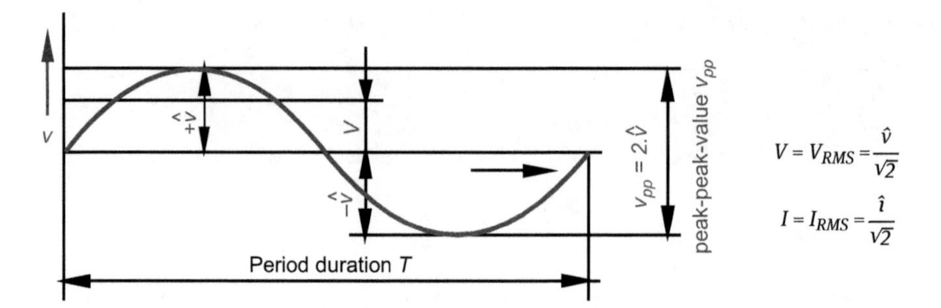

Figure 7.5: Peak value and effective value for sinuisoidal quantities.

$V = V_{RMS}$	Effective value of voltage
$I = I_{RMS}$	Effective value of current
$\hat{v}$	Peak-value of voltage
v_{pp}	Peak-peak value

7.4 Powers in the alternating-current circuit

As an introduction, we start with a memory aid to differentiate between the respective types of power (see Figure 7.6). The geometrical sum of active and reactive power is called apparent power.

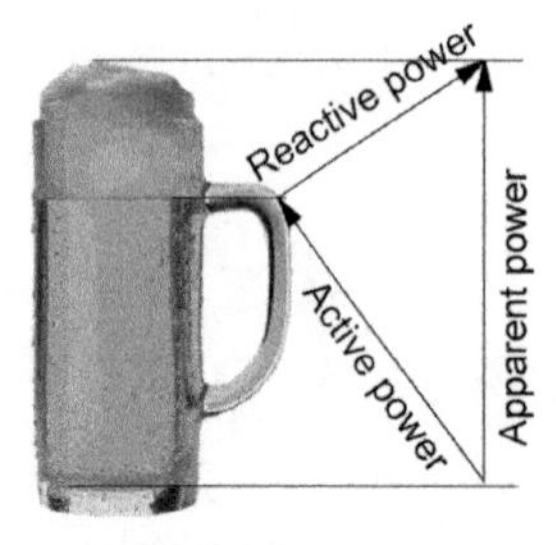

Figure 7.6: Memory aid regarding the terms apparent power, active power and reactive power.

7.4.1 Active power P

If an active resistor (e.g. heating resistor) is operated in an alternating-current circuit, voltage and current are in phase. The active power P is the arithmetic mean resulting from the multiplication of instantaneous values of current $i(t)$ and voltage $v(t)$ or from the product of the effective values of current and voltage:

For $\varphi = 0°$ the following applies:

$$P = \frac{1}{2} \cdot \hat{p} = \frac{1}{2} \cdot \hat{v} \cdot \hat{\imath} = \frac{1}{2} \cdot \sqrt{2} \,.\, V \,.\sqrt{2} \,.\, I$$
$$= U \,.\, I [P] = W$$

Generally (arbitrary phase angle) the following applies:

$$\mathbf{P = V \,.\, I \,.\, \underbrace{\cos(\varphi)}_{=1 \text{ for } \varphi = 0°}}$$

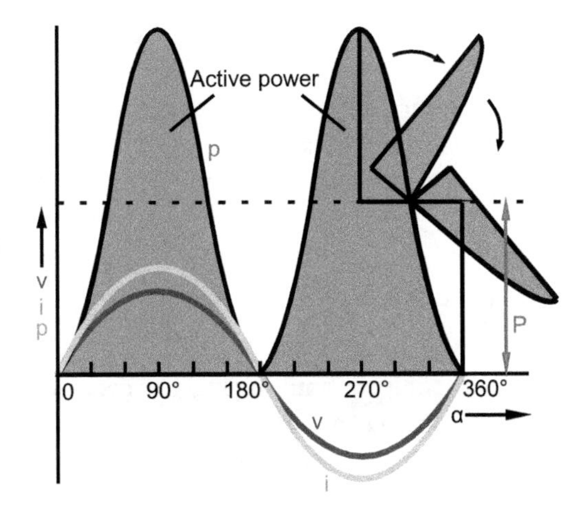

Figure 7.7: AC power with active load.

7.4.2 Reactive power *Q*

In an alternating-current circuit with an inductive or capacitive resistor,[84] the phase shift between current and voltage is 90°. The number of positive areas of the power curve equals the number of negative ones. Consequently, the active power is zero and only reactive power occurs. The total amount of energy alternates between the load and the generator; it is not converted into another form of energy, e.g. heat. Reactive power is a power that does not produce an effect.

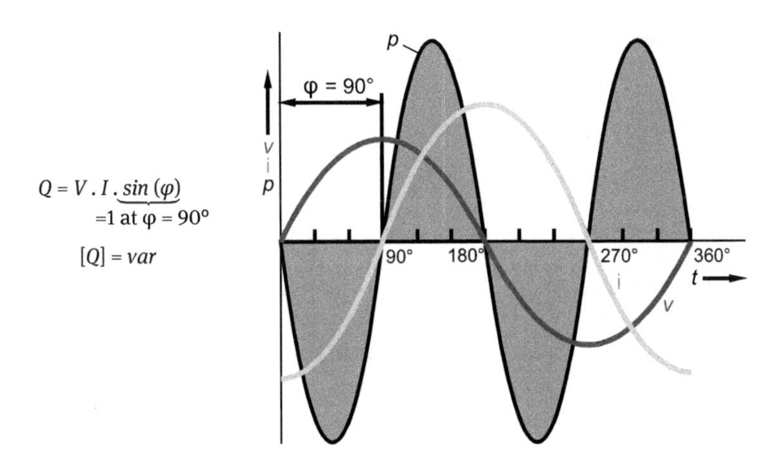

Figure 7.8: Inductive reactive power.

84 ... This is however impossible in real life as every coil and every capacitor also has an ohmic resistance.

Inductive reactive power Q is needed to form magnetic fields, oscillating between grid and coil.

Due to reactive power, electrical lines[85] need to be built for a higher current which leads to higher costs. Therefore, the electricity supplier requires industrial clients to ensure reactive-current compensation.[86]

7.4.3 Apparent power *S*

If an active resistor R and an inductive resistor (X_L) or a capacitive reactance (X_C) are combined,[87] multiplying the measured values of voltage and current we obtain an apparent power S. The active power[88] however is lower – it equals the arithmetic mean over a period; the reactive power is cancelled out.

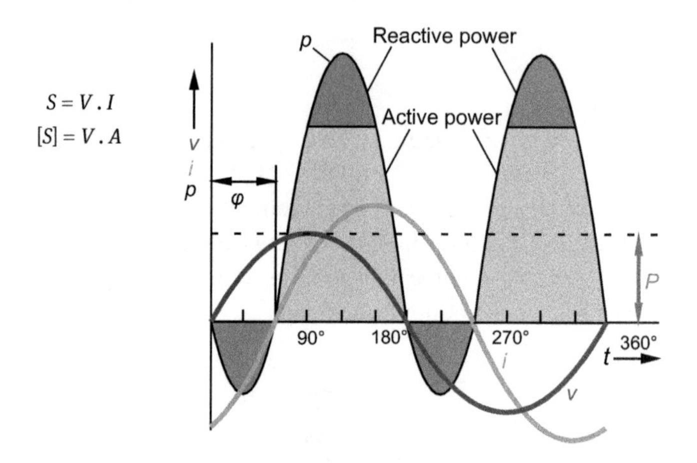

Figure 7.9: AC power (here: φ = 60°).

7.4.4 Correlation between *S*, *P* and *Q*

The correlation between the three types of power can be represented with a right-angled triangle (=power triangle). In a series connection with an active resistance R and an inductive reactance X_L, the power triangle is similar to the voltage triangle

85 … Here we refer to the electrical lines from the generator to the load.

86 … To compensate the reactive power, capacitors are used: "reactive-current compensation".

87 … which is generally the case.

88 … which, in this example, refers to the power that is converted into electric heat at the active resistor.

as in the power equations $S = V \cdot I$ and $P = V_A \cdot I$ and $Q_L = V_{rL} \cdot I$ always the same current is applied.

In a parallel connection with an active resistance and an inductive reactance, the power triangle is similar to the current triangle.

The angle between P and S equals the phase shift angle φ.

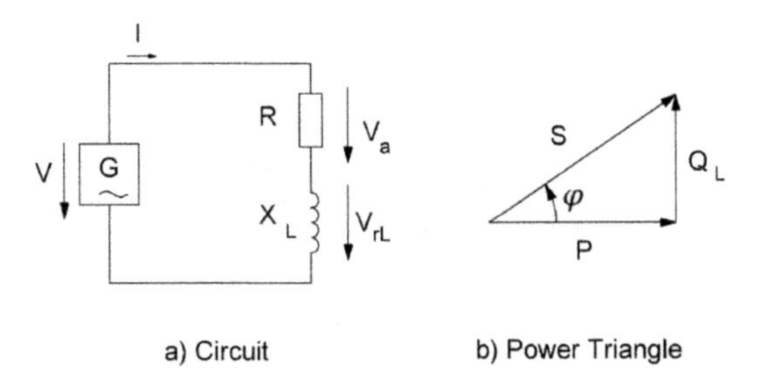

Figure 7.10: Powers in an R-X_L series connection.

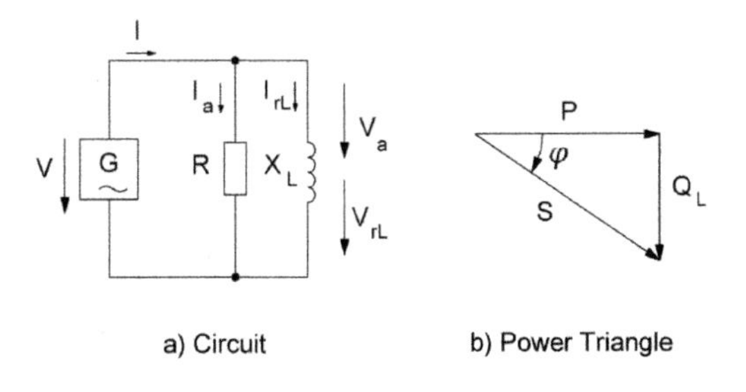

Figure 7.11: Powers in an R-X_L parallel connection.

Table 7.2: Power with inductive load.

$S^2 = P^2 + Q_L^2$	$\Rightarrow S = \sqrt{P^2 + Q_L^2}$	$S = V \cdot I$
$\cos(\varphi) = \frac{P}{S}$	$\Rightarrow P = S \cdot \cos(\varphi)$	$P = V \cdot I \cdot \cos(\varphi)$
$\sin(\varphi) = \frac{Q_L}{S}$	$\Rightarrow Q_L = S \cdot \sin(\varphi)$	$Q_L = V \cdot I \cdot \sin(\varphi)$
$\tan(\varphi) = \frac{Q_L}{P}$		$Q_L = P \cdot \tan(\varphi)$

Active power factor cos (φ) and power factor λ

The ratio of active power P to complex power S is called active power factor $\cos(\varphi)$. It is indicated e.g. at the power rating of an AC motor.

$$\boldsymbol{\cos(\varphi) = \frac{P}{S}}$$

The power factor λ is calculated as follows:

$$\lambda = \frac{P}{S}$$

7.5 R, X_L and X_C in the alternating-current circuit

7.5.1 Active resistance R

A resistor that has the same effect in the alternating-current circuit as well as in the direct-current circuit is called active resistor *R*. At the active resistor, voltage and current are in phase.

In the **active resistor *R*** electrical energy is converted into another form of energy (e.g. light bulbs, heaters, semiconductor resistors).

The active resistance is calculated using the effective values:

$$R = \frac{V}{I}$$

$$[R] = \frac{V}{A} = \Omega$$

7.5.2 Electrical impedance *Z*

The resistance of the coil is higher with sinusoidal voltage than with direct voltage. The resistance with sinusoidal current is called electrical impedance Z (impedance[89]).

The **impedance *Z*** can be calculated using the effective values of sinusoidal current *I* and sinusoidal voltage *V*:

$$Z = \frac{V}{I}$$

$$[Z] = \frac{V}{A} = \Omega$$

Current and voltage are not in phase.

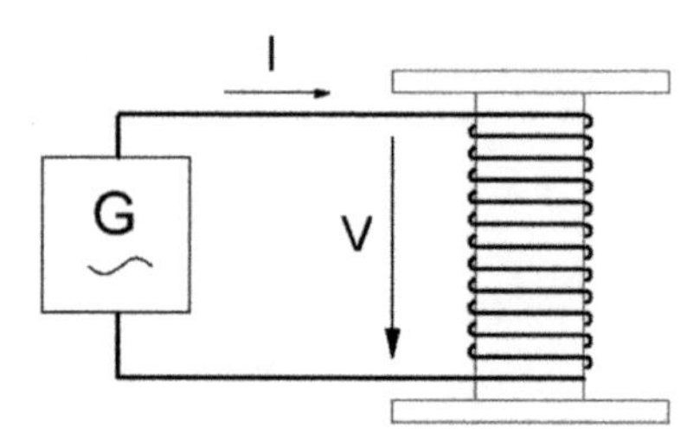

Figure 7.12: Impedance of a coil.

89 derived from impedire (lat.) = to hinder, inhibit.

7.5.3 Inductive reactance X_L in the alternating-current circuit

The current consumption of a coil with alternating voltage (see Figure 7.13) is lower than with direct voltage. The cause of this is the inductive reactance. The inductive reactance is generated by the self-induced voltage in the coil which inhibits the sinusoidal current. The current generates a magnetic flux Φ in phase. As a result of the flux alternation, a self-induction voltage is created which – in accordance with Lenz's law – has an inhibiting effect on the increase or decrease of the current at any given moment. The coil current reaches its peak value a quarter period after the self-induction voltage.

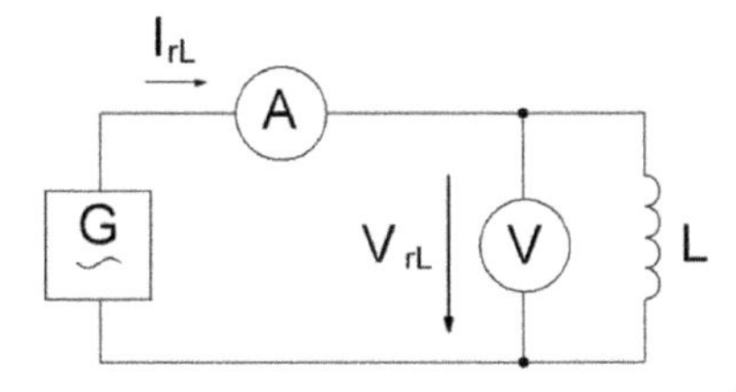

Figure 7.13: Alternating voltage applied to a coil.

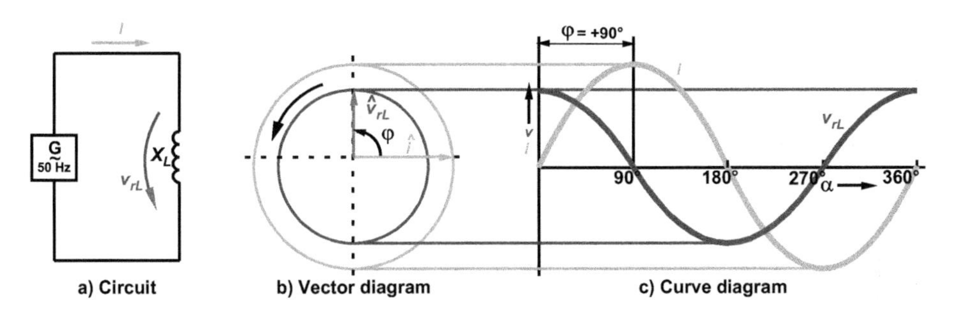

Figure 7.14: Current and voltage at an inductive reactance (R=0; ideal coil).

With inductive reactance, the **sinusoidal current lags** the sinusoidal voltage by **90°.** The higher the inductance L of a coil and the higher the (angular) frequency, the higher is the inductive reactance X_L of the coil.

$$X_L = \frac{V}{I} = \underbrace{\omega}_{2\cdot\pi\cdot f} \cdot L$$

X_L Inductive reactance in Ω

I Current in A

ω Angular frequency: $\omega = 2\cdot\pi\cdot f$

L Inductance; $[L] = \frac{Vs}{A} = H$ (Henry)

V_{rL} Inductive reactive voltage in V

7.5.4 R and X_L in the alternating-current circuit

Series connection

Let us have a look at the series connection of a choke coil and an active resistor and its equivalent circuit (see Figure 7.15).

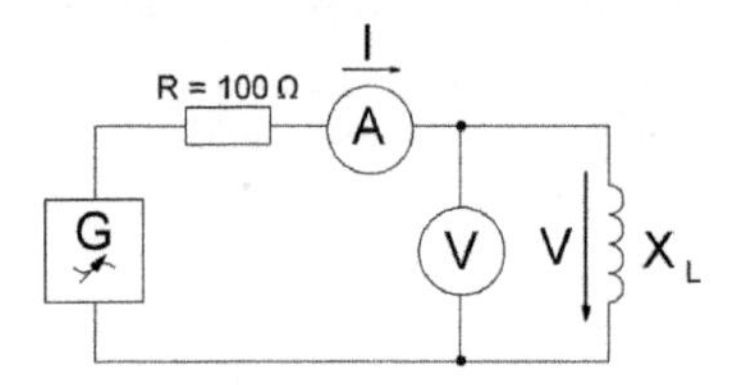

Figure 7.15: Phase shift caused by a coil.

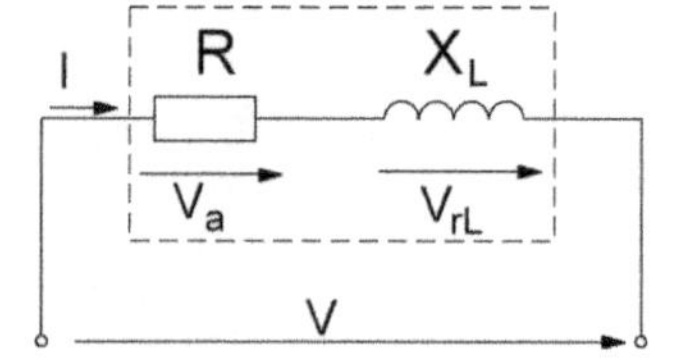

Figure 7.16: Equivalent circuit of a coil.

In a real coil, the phase shift between current and voltage is always **smaller than 90°** due to the active resistance of the coil. At the active resistor R, the active voltage V_a (in phase) drops; at the inductive reactance X_L the inductive reactive voltage V_{rL} drops (it is ahead of the current by 90°). The peak value $\hat{v}$ is smaller than the sum of $\hat{v}_a$ and $\hat{v}_{rL}$. The total voltage is ahead of the shared current by the phase shift angle φ.

Approach on how to draw a vector diagram (see Figure 7.17).

1. Determine the scale
2. Reference value on horizontal axis
3. Active voltage and current are in phase
4. Inductive reactive voltage is ahead of the current by 90°
5. Total voltage calculated through vector addition

The total voltage is ahead of the shared current by the phase shift angle φ.

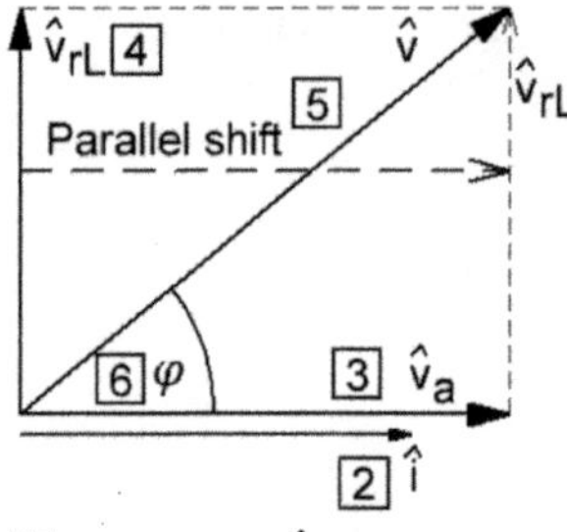

Figure 7.17: Vector diagram.

We notice that in the vector diagram the reactive voltage and the active voltage are perpendicular to one another and that the total voltage is ahead of the shared current by a certain angle φ that is determined by the level of reactive voltage. In order to simplify the vector diagram, we use the so-called voltage triangle in which we insert the effective values of the voltages:

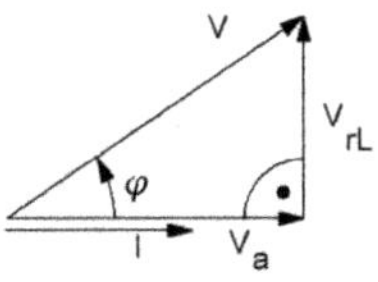

In the **voltage triangle** the following applies:

$$V = \sqrt{V_a^2 + V_{rL}^2}$$

$$\sin(\varphi) = \frac{V_{rL}}{V} \Rightarrow V_{rL} = V \cdot \sin(\varphi)$$

$$\cos(\varphi) = \frac{V_a}{V} \Rightarrow V_a = V \cdot \cos(\varphi)$$

Figure 7.18: Voltage triangle.

To determine the resistances, a resistance triangle is applied:

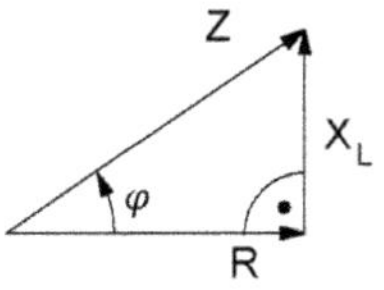

In the **resistance triangle** the following applies:

$$Z = \frac{V}{I} = \sqrt{R^2 + X_L^2}$$

$$X_L = \frac{V_{rL}}{I} = Z \cdot \sin(\varphi)$$

$$R = \frac{V_a}{I} = Z \cdot \cos(\varphi)$$

Figure 7.19: Resistance triangle.

Parallel connection

Let us have a look at this parallel connection with an active resistor and an inductive reactor (Figure 7.20 a). The corresponding vector and curve diagram are represented in Figure 7.20 b and Figure 7.20 c.

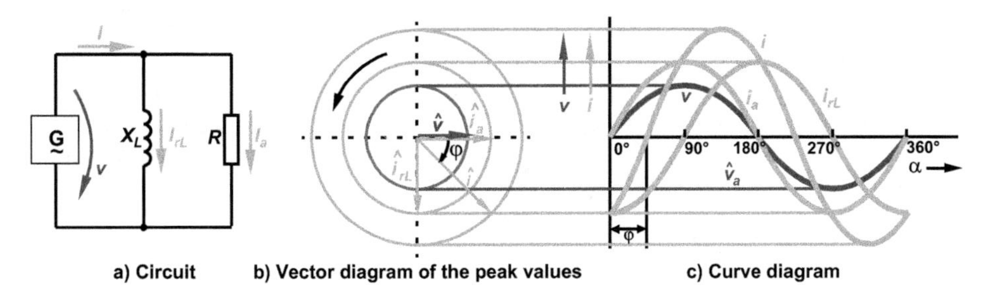

Figure 7.20: Parallel connection with active resistor and inductive reactor.

The total current I is divided into a current I_{rL} which flows through the inductive reactor X_L and a current I_a through the active resistor R. At both resistors, the shared voltage V is applied. The active current I_a is in phase with the shared voltage V. The reactive current I_{rL} lags the shared voltage **by 90°**. The total current I is the arithmetic sum (current triangle) of active current I_a and reactive current I_{rL}. It lags the voltage V by the phase shift angle φ. In the parallel connection, we use the current triangle to calculate the total current I:

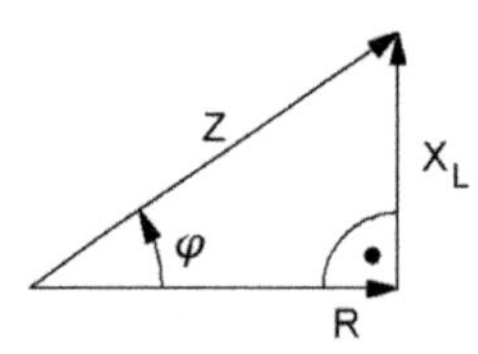

In the **current triangle** the following applies:

$$I = \sqrt{I_a^2 + I_{rL}^2}$$

$$I_{rL} = I \,.\, \sin(\varphi)$$

$$I_a = I \,.\, \cos(\varphi)$$

Figure 7.21: Current triangle.

The impedance is calculated by adding up the conductances. To do so, we use the conductance triangle which is analogous to the resistance triangle:

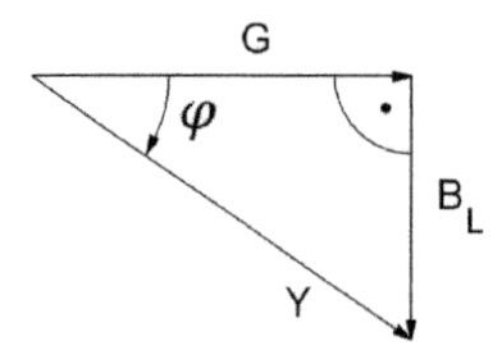

In the **conductance triangle** the following applies:

$$Y = \frac{1}{Z} = \sqrt{G^2 + B_L^2}$$

$$Y = \frac{1}{Z} \qquad G = \frac{1}{R} \qquad B_L = \frac{1}{X_L}$$

$$[Y] = \frac{1}{\Omega} \qquad [G] = \frac{1}{\Omega} = S\,(Siemens) \qquad [B_L] = \frac{1}{\Omega}$$

Figure 7.22: Conductance triangle.

7.5.5 Capacitive reactance X_C in the alternating-current circuit

Taking a look at the sinusoidal voltage of an (ideal[90]) capacitor (Figure 7.23), we notice that the capacitor is constantly charged and recharged. At the capacitor, a sinusoidal voltage V_{rC} with the same frequency as the sinusoidal current is generated. The capacitor, hence, also is a resistor and referred to as capacitive reactor X_C.

$$X_C = \frac{1}{\omega \cdot C} = \frac{1}{2\pi \cdot f \cdot C}$$

90 without loss, i.e. $R = 0\Omega$.

X_C Capacitive reactance; $[X_C] = \Omega$
ω Angular frequency: $\omega = 2 \cdot \pi \cdot f$
C Capacitance; $[C] = \frac{As}{V} = F$ (Farad)

At the capacitive reactor X_C, **the sinusoidal current is ahead** of the sinusoidal voltage **by 90°**.

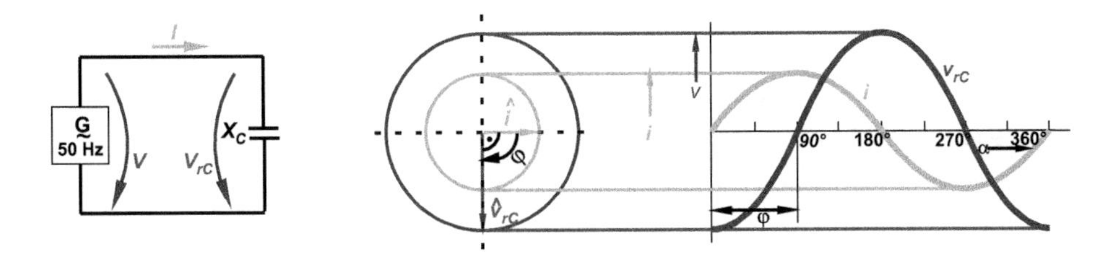

Figure 7.23: Current and voltage at a capacitive reactor.

7.5.6 R and X_C in the alternating-current circuit

Series connection

When connecting an active resistor R and a capacitive reactor X_C in series to a voltage (see Figure 7.24) with a sinusoidal temporal sequence, the active voltage V_a is in phase with the current I; the capacitive reactive voltage V_{rC} lags the current by 90°. Therefore, in the voltage triangle, the vector of the reactive voltage V_{rC} is represented as lagging 90° behind compared to the vector of the shared current I.

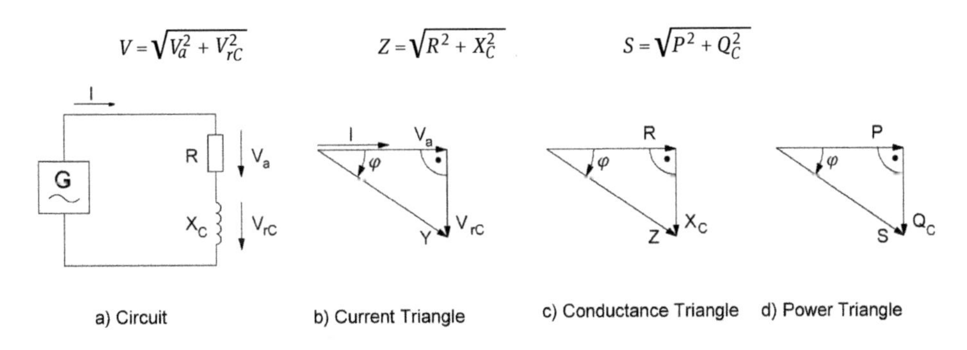

Figure 7.24: Series connection with the active resistor R and capacitive reactor X_C.

The resistance triangle or power triangle of a series connection can be derived from the voltage triangle dividing the vectors of the voltages by the shared current or multiplying said vectors by the voltage. Therefore, the triangles are similar. The quantities of the vector triangles can be calculated using the Pythagorean theorem or the trigonometric functions.

Parallel connection

If a capacitive reactor X_C and an active resistor R are connected in parallel (see Figure 7.25 a), the total current divides up into a capacitive reactive current I_{rC} and an active current I_a. The capacitive reactive current I_{rC} is ahead of the shared voltage V by 90°. In parallel connections, we use the current triangle (see Figure 7.25 b).

The conductances (see Figure 7.25 c) and powers (see Figure 7.25 c) can again be calculated using the Pythagorean theorem or the trigonometric functions:

Conductances	Power
$Y=\sqrt{G^2+B_C^2}$	$S=\sqrt{P^2+Q_C^2}$
$\cos(\varphi)=\frac{G}{Y}$ $\sin(\varphi)=\frac{B_C}{Y}$	$\cos(\varphi)=\frac{P}{S}$ $\sin(\varphi)=\frac{Q_C}{S}$
$Y=\frac{1}{V}=\frac{1}{Z}$ $G=\frac{I_a}{V}=\frac{1}{R}$ $B_C=\frac{I_{rC}}{V}=\frac{1}{X_C}$	$S=V\cdot I$ $P=V\cdot I_a$ $Q_C=V\cdot I_{rC}$
Y Admittance in S(Siemens) G Conductance in S(Siemens) B_C Susceptance in S(Siemens) I Total current in A V Shared voltage in V Z Impedance in Ω I_a Active current in A R Active resistance in Ω	I_{rC} Reactive current in A X_C Capacitive reactance in Ω S Apparent power in VA P Active power in W Q_C Capacitive reactive power in var

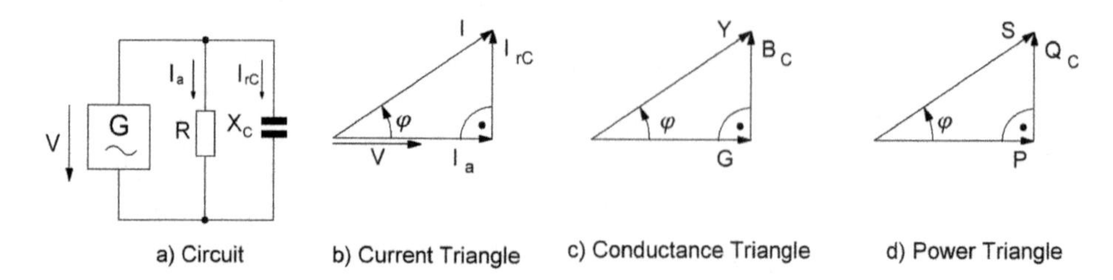

Figure 7.25: Parallel connection of R and X_C.

7.5.7 R, X_L and X_C in an alternating-current circuit

Series connection

In series connection, the capacitive and the inductive reactive voltage work against each other as the inductive reactive voltage V_{rL} **is ahead of the** current by

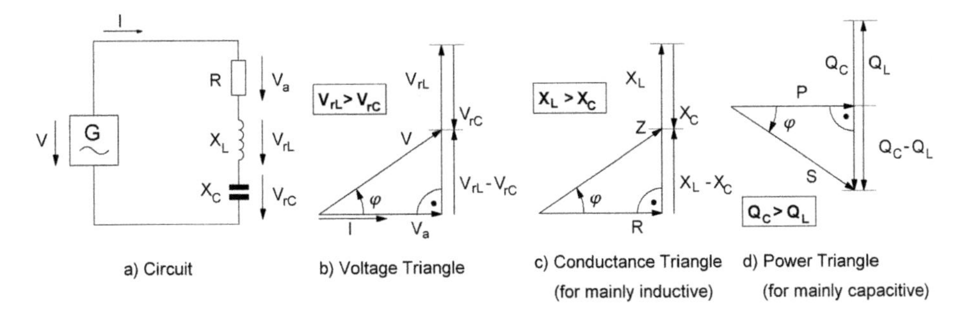

Figure 7.26: Series connection with R,, X_L and X_C.

90° while the capacitive reactive voltage V_{rC} lags the current by 90°. Therefore, also the vectors representing the reactive voltages in the vector diagram oppose each other. The total voltage is calculated through vector addition of the partial voltages.

Voltages in the series connection: $V = \sqrt{V_a^2 + (V_{rL} - V_{rC})^2}$

Resistances and powers:

$$Z = \sqrt{R^2 + (X_L - X_C)^2} \qquad Z = \frac{V}{I}$$

$$S = \sqrt{P^2 + (Q_C - Q_L)^2} \qquad S = V \cdot I$$

$$P = I^2 \cdot R \qquad Q_L = I^2 \cdot X_L \qquad Q_C = I^2 \cdot X_C$$

Parallel connection

In the parallel connection with R, X_L and X_C (see Figure 7.27 a) the current I_{rL} is behind the shared voltage V by 90°. Due to the capacitance, the current I_{rC} is ahead of the voltage V by 90° (see Figure 7.27 b). Hence, I_{rC} and I_{rL} always oppose each other. Consequently, the capacitance acts as a load in the time periods in which the inductance acts as a generator. The total current is the sum of the vector addition of the partial currents. If the inductive reactive current I_{rL} is higher than the capacitive reactive current I_{rC}, the parallel connection with R, X_L and X_C predominantly acts inductively. In case the capacitive reactive current I_{rC} is higher than the inductive reactive current I_{rC}, the circuit predominantly acts capacitively. The vector diagram of conductances is the result of dividing the vectors in the current triangle by the voltage V. By multiplying the vectors in the current triangle with the voltage V, we derive at the vector diagram of the powers.

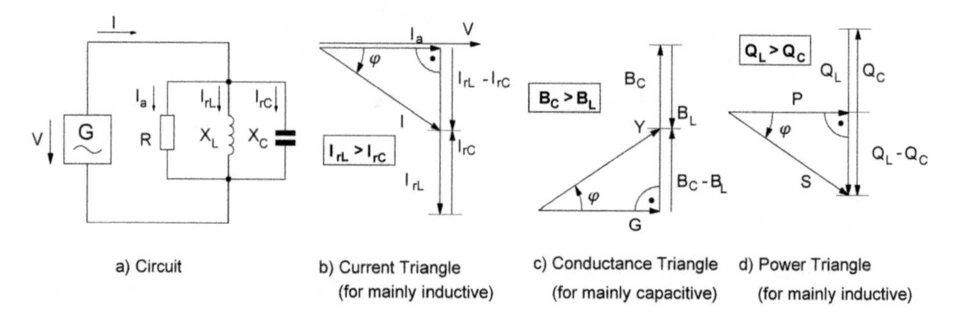

Figure 7.27: Parallel connection with R, XL and XC.

Currents in the parallel connection

$$I = \sqrt{I_a^2 + (I_{rL} - I_{rC})^2}$$

Conductances and powers

$$Y = \frac{1}{Z} = \sqrt{G^2 + (B_C - B_L)^2}$$

$$S = V \cdot I$$

$$P = V \cdot I_a \qquad Q_L = V \cdot I_{rL} \qquad Q_C = V \cdot I_{rC}$$

7.6 Resonant circuits

A resonant circuit consisting of a coil and a capacitor is used to either filter out or suppress a specific frequency (resonance frequency) in a large number of signals. The charged capacitor discharges via the coil. The discharging current in the coil creates a magnetic field. After the capacitor is fully discharged, the magnetic field is dissipated. The alteration in the magnetic field induces a voltage in the coil. The capacitor is charged in reverse polarity until the magnetic field in the coil is completely dissipated. The capacitor voltage generates an electric field in the capacitor. As a consequence, the current now creates a magnetic field in the coil. Electric and magnetic fields occur in succession (see Figure 7.28). This process repeats itself periodically.

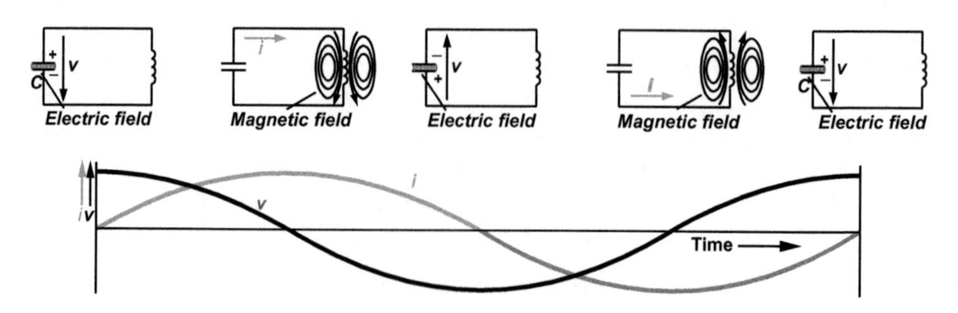

Figure 7.28: Electric and magnetic field of a resonant circuit.

In the resonant circuit, a periodical exchange of electrical energy in the capacitor and of magnetic energy in the coil and vice versa happens.

The alternating current in the resonant circuit generates heat, especially through the active resistance in the coil. This withdraws energy from the system and, over time, leads to decreasing amplitude of the oscillation (see Figure 7.29). If the resonant circuit is externally resupplied with energy of adequate frequency, it starts the transient phase (see Figure 7.30).

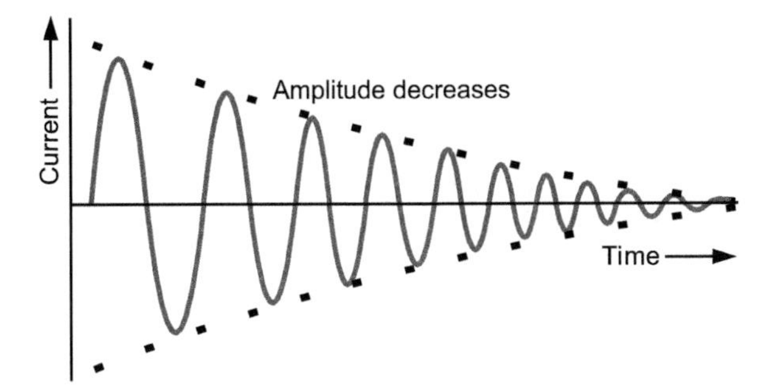

Figure 7.29: Decreasing amplitude.

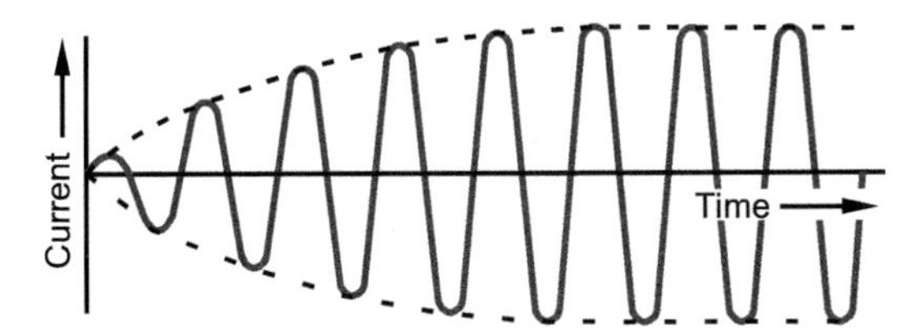

Figure 7.30: Transient phase.

Resonance

The inductance and the capacitance determine the natural frequency of the resonant circuit. This is the frequency at which the energy alternates between the capacitor and the coil.

7.6.1 Series resonant circuit

In a series resonant circuit, coil and capacitor are connected in series (see Figure 7.31). In the series resonant circuit, the alternating current reaches its maximum at resonance. A voltage increase (voltage resonance) generally occurs at the coil and at the capacitor. The reactive voltages V_{rL} and V_{rC} can exceed the total voltage by many times.

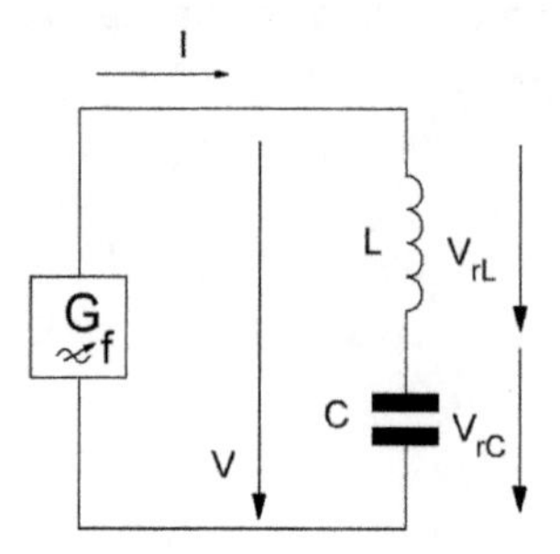

$V_{rL} = V_{rC} \Rightarrow$ resonance

Figure 7.31: Series resonant circuit.

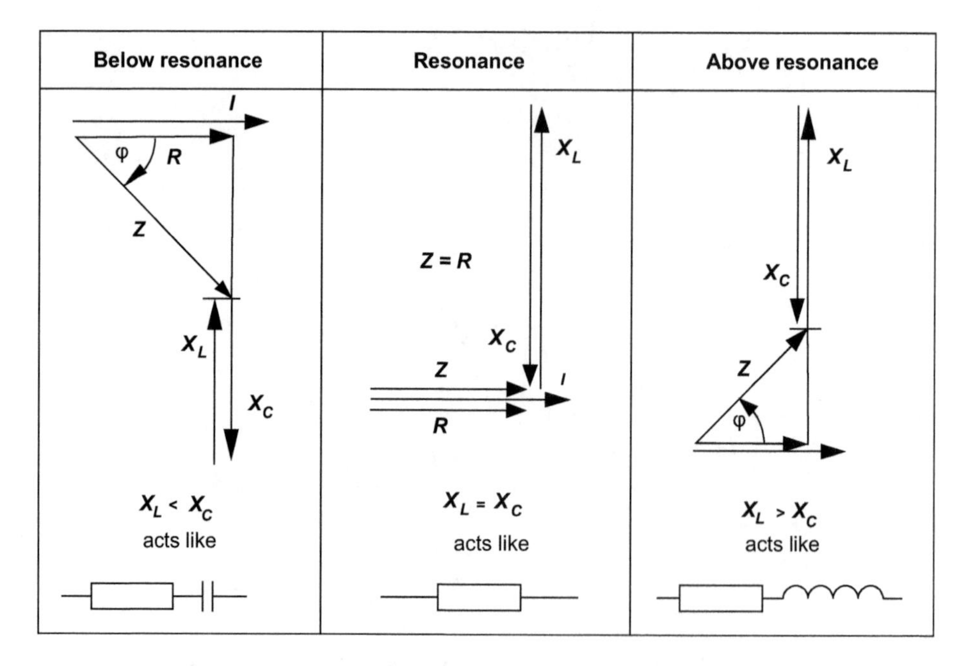

Figure 7.32: Vector diagram of the series resonant circuit.

Resonance condition

At resonance, the voltages at the coil and at the capacitor are equal (see Figure 7.32).

The inductive reactance X_L then equals the capacitive reactance X_C.

$$\Rightarrow \omega_r \cdot L = \frac{1}{\omega_r \cdot C}$$

$$\Rightarrow \omega_r^2 = \frac{1}{L \cdot C}$$

$$\Rightarrow f_r = \frac{1}{2\pi \cdot \sqrt{L \cdot C}}$$

f_r Resonance frequency in Hz
R_r Resonance resistance in Ω
R_l Loss resistance of the coil in Ω
L Inductance in H
C Capacitance in F

Below the resonance frequency, the capacitive resistance prevails in the series resonant circuit; above the resonance frequency, the inductive resistance is dominant. If we draw the development of the impedance Z depending on the frequency f, we obtain the resonance curve of the series resonant circuit (see Figure 7.33). With a series resonant circuit, we can short-circuit the resonance frequency with a frequency mixture.

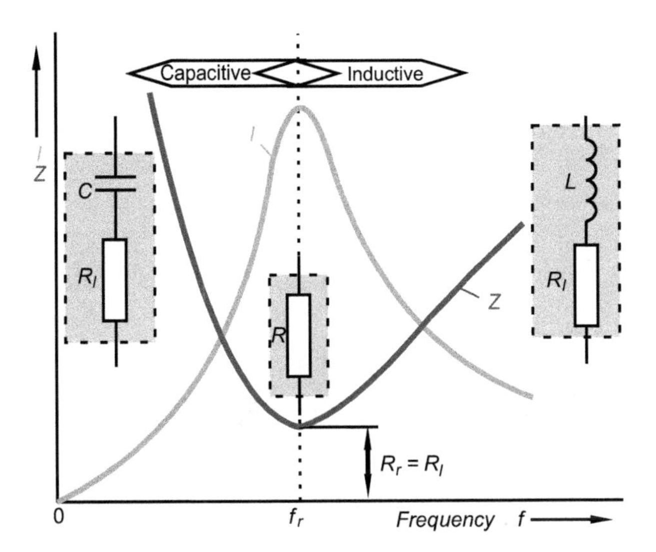

Figure 7.33: Resonance curve of the series resonant circuit.

7.6.2 Parallel resonant circuit

In a parallel resonant circuit, coil and capacitor are connected in parallel. At resonance, the partial currents are equal due to inductance and capacitance. The inductive reactance is as high as the capacitive reactance because the shared voltage is applied to both, coil and capacitor.

In the parallel resonant circuit, the total current in the supply is the lowest at resonance.

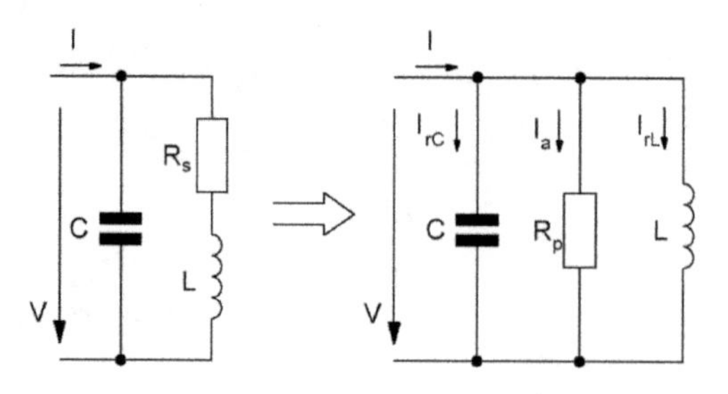

Figure 7.34: Equivalent schematic of the parallel resonant circuit.

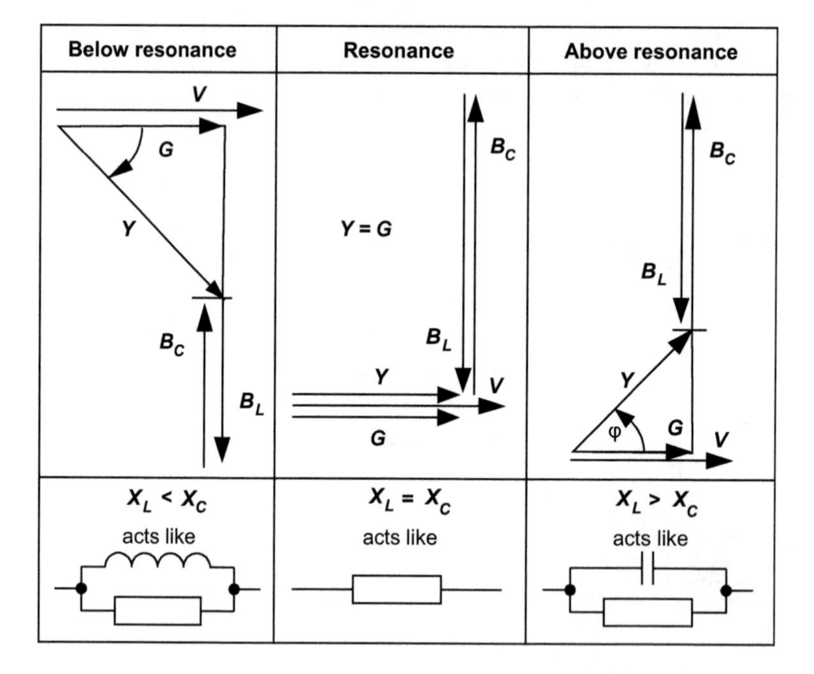

Figure 7.35: Vector diagram of the parallel resonant circuit (for I = const.).

At resonance, $R_r = R_p$ with $R_p \approx L/C \cdot R_l$

The resonance frequency f_r is calculated as follows:

$$f_r = \frac{1}{2\pi \cdot \sqrt{L \cdot C}}$$

f_r Resonance frequency in Hz
L Inductance in H
C Capacitance in F

The parallel resonant circuit is used to filter out a specific frequency – the resonance frequency – from a frequency mixture.

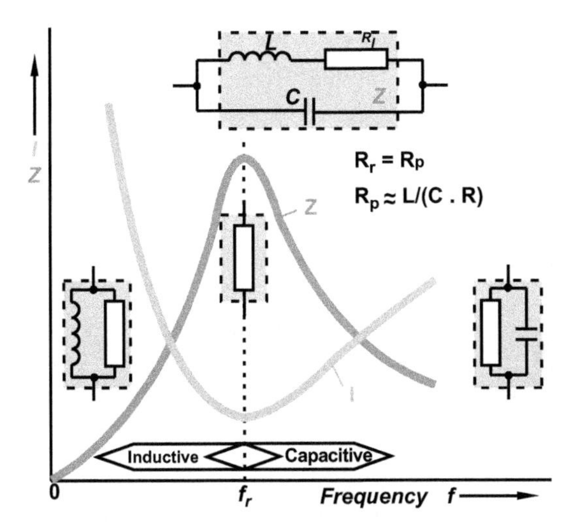

Figure 7.36: Parallel resonant circuit.

Resonance condition

At resonance, the partial currents are equal due to inductance and capacitance. The inductive reactance is as high as the capacitive reactance because the shared voltage is applied to both, coil and capacitor. Therefore, the same resonance condition as in a series resonant circuit applies.

$$X_L = X_C$$

$$\Rightarrow \omega_r \cdot L = \frac{1}{\omega_r \cdot C}$$

$$\Rightarrow \omega_r^2 = \frac{1}{L \cdot C}$$

$$\Rightarrow f_r = \frac{1}{2\pi \cdot \sqrt{L \cdot C}}$$

7.7 Harmonics, Fourier series representation

In electrical engineering, sinusoidal oscillations as well as sinusoidal voltage and current curves are of great importance. In a power plant, sinusoidal voltages are generated. These would usually create sinusoidal currents. Electrical appliances (coils filled with iron at magnetic saturation, rectifiers, phase-fired controls etc.), however, distort the sinusoidal shape. That way, the current is still periodic but no longer sinusoidal. We say that “the oscillation has harmonics”.

Each periodic non-sinusoidal oscillation can be perceived as composition of several sinusoidal parts into which it can be decomposed again.

Each arbitrary non-sinusoidal periodic function $f(t)$ can thus be depicted by a Fourier series in the form of:

$$f(t) = A_0 + \sum_{n=1}^{\infty} C_n \cdot \sin(n + \omega_1 t + \varphi_n)$$

A_0 Constant (temporary mean, DC coefficient)
T Time period of the non-sinusoidal ouput function
ω_1 Lowest occurring angular frequency; $\omega_1 = 2\pi \cdot f_1$ ($f_1 \ldots$ lowest occurring frequency)
C_n Constant with the peak value that corresponds to the nth harmonic
φ_n Related zero phase angle

Each periodic non-sinusoidal oscillation can be seen as a composition of several sinusoidal and cosine parts into which it can be decomposed again.

Each periodic non-sinusoidal oscillation can be indicated by decomposition into cosine and sine components as follows:

$$f(t) = A_0 + \sum_{n=1}^{\infty} A_n \cdot \cos(n\omega_1 t) + B_n \cdot \sin(n\omega_1 t)$$

$$C_n = \sqrt{A_n^2 + B_n^2} \qquad \varphi_n = \arctan\left(\frac{A_n}{B_n}\right)$$

The distortion factor **d** indicates the harmonic content of a periodic process:

$$d = \frac{\textit{Effective value of the harmonic}}{\textit{Effective value of the total oscillation}}$$

7.7.1 Generation of a square wave through overlap of sinusoidal oscillations (Fourier synthesis)

Reversely to the Fourier analysis during which we look at the frequency spectrum[91] of a periodic signal, we can produce a desired signal through targeted overlap of sinusoidal oscillations with corresponding frequency and amplitude (Fourier synthesis).

---

91 The frequency spectrum includes all frequencies of the harmonics.

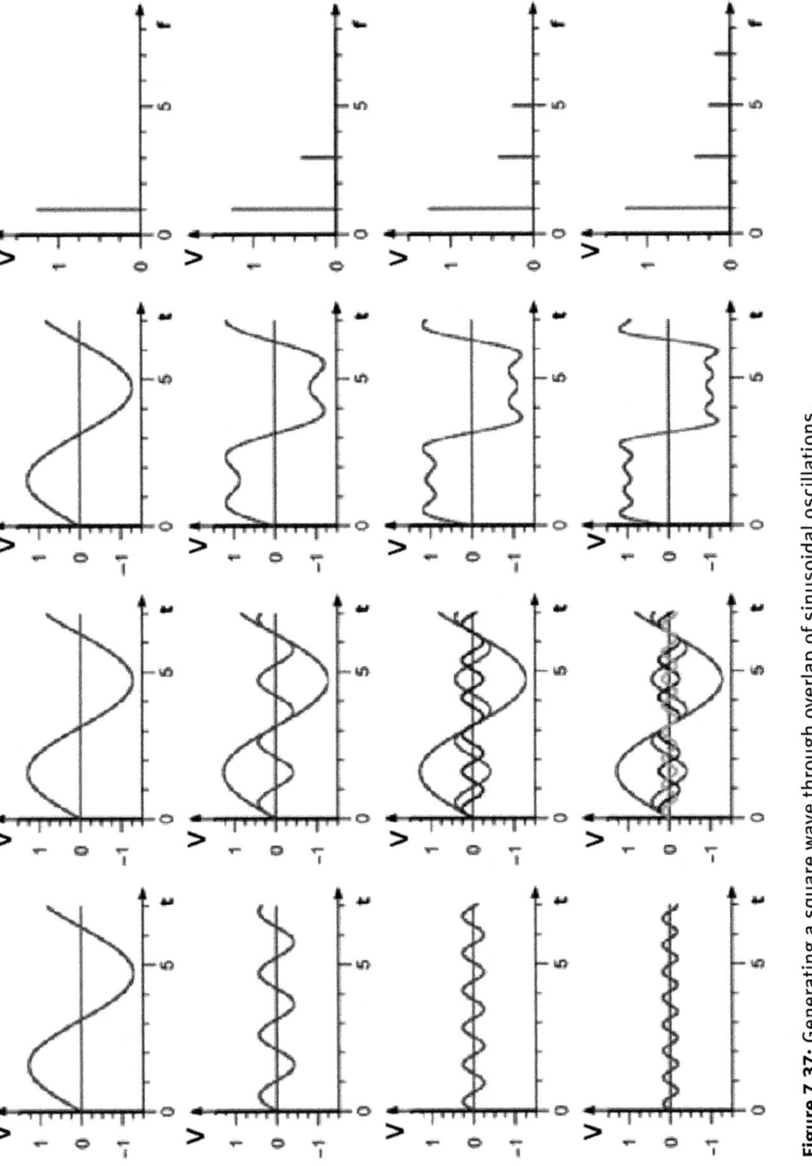

Figure 7.37: Generating a square wave through overlap of sinusoidal oscillations.

If, for example, a square wave is desired, it can be represented with overlaps of sinusoidal oscillations. If it is only a sinusoidal oscillation (see Figure 7.37, first line, first diagram) the result (see Figure 7.37, first line, third diagram) does not resemble a square wave. If we overlap this so-called fundamental frequency with a sine wave with triple frequency (see Figure 7.37, second line) "first harmonic wave", the result already resembles a square wave a bit more. Adding more harmonic waves constantly improves the result (e.g. Figure 7.37, fourth line, third diagram). The frequencies of these harmonic waves are evident in the respective frequency spectrum (see Figure 7.37, first line, diagrams to the far right).

7.8 Three-phase current (rotary current)

Rotary current is the customary name for three-phase current. It is defined as a system of three alternating-current circuits with their respective voltages which temporally lag by a third of a period with respect to each other. Rotary current is produced in three-phase generators through induction in three windings displaced by 120° (see Figure 7.38). For the long-distance transmission of electric power, the beginnings of the windings are interconnected to a neutral point and we obtain a so-called "neutral conductor". That way, we only need 4 lines instead of 6 lines for three-phase current. When the three-phase supply is loaded symmetrically, the neutral conductor can be omitted, because all currents at each moment would sum up to zero. It is possible to transmit a three-phase system with only three lines. That way, only half of the conductors are necessary.[92]

If coils, displaced by 120°, are connected to these 3 voltages, we get a rotating magnetic field due to the flowing currents. This is the prerequisite for the rotation of the rotors in (e.g. asynchronous) motors.

The breakthrough for rotary current began in 1891 with the successful transmission of rotary current from the hydroelectric power plant Lauffen am Neckar to Frankfurt am Main, 175 km away. A power of 165 kW with an efficiency of about 75 % was transmitted. Before that, most electrical engineering pioneers like Edison used direct current.[93] Rotary current is usually produced by synchronous inner pole generators.

---

92 On high-voltage-line towers there are often 7 lines: the top conductor is the earthed lightning protection rope (protects the other conductors from lightning strikes) and 2 three-phase electric systems (with 3 conductors, respectively).

93 The so-called "War of the Currents" between Thomas Alva Edison and George Westinghouse was an economic discussion concerning the issue which one would be the more suitable technology to provide an extensive electrical energy supply as well as to develop a power grid for the USA. Edison favoured the direct current technology, but he was defeated by Westinghouse who favoured the three-phase AC technology.

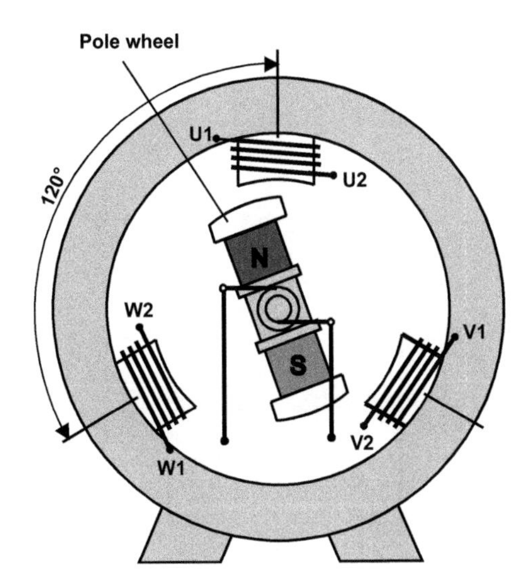

Figure 7.38: Generation of three-phase voltage.

If the pole wheel rotates, alternating voltage with the same amplitude and frequency is induced in each coil. The voltages are temporally shifted by one third of a period due to their layout. The phase shift angle is 120°, respectively. The three coils of such a generator form the strings of the machine. In each string, a voltage called string voltage is induced. The beginnings of the strings are called U1, V1, W1 and the ends of the strings are called U2, V2, W2. Through linking (connecting) the coils, the number of conductors needed for energy transfer can be reduced to three (C1, C2, C3).

7.8.1 Concatenation

If the three ends of strings U2, V2 and W2 are connected at the generator or the load, a wye connection occurs (symbol: Y). The point of connection of U2, V2 and W2 is called neutral point. Usually the **neutral conductor N** is connected at the neutral point.

When the end of a string is connected with the beginning of the next one, e.g. U2 with V1, V2 with W1 and W2 with U1, a delta connection with the symbol is the result. The three conductors C1, C2 and C3 that in the circuits lead from the generator to the beginnings of the strings U1, V1 and W1 are called **exterior conductors.** If we form the quotient of the voltage between two exterior conductors (e.g. $V_{31} = 400\ V$) and the voltage between exterior conductor and neutral conductor N, e.g. $V_{1N} = 230\ V$, we obtain the **concatenation factor** $\sqrt{3}$:

$$\frac{400\ V}{230\ V} = \sqrt{3}$$

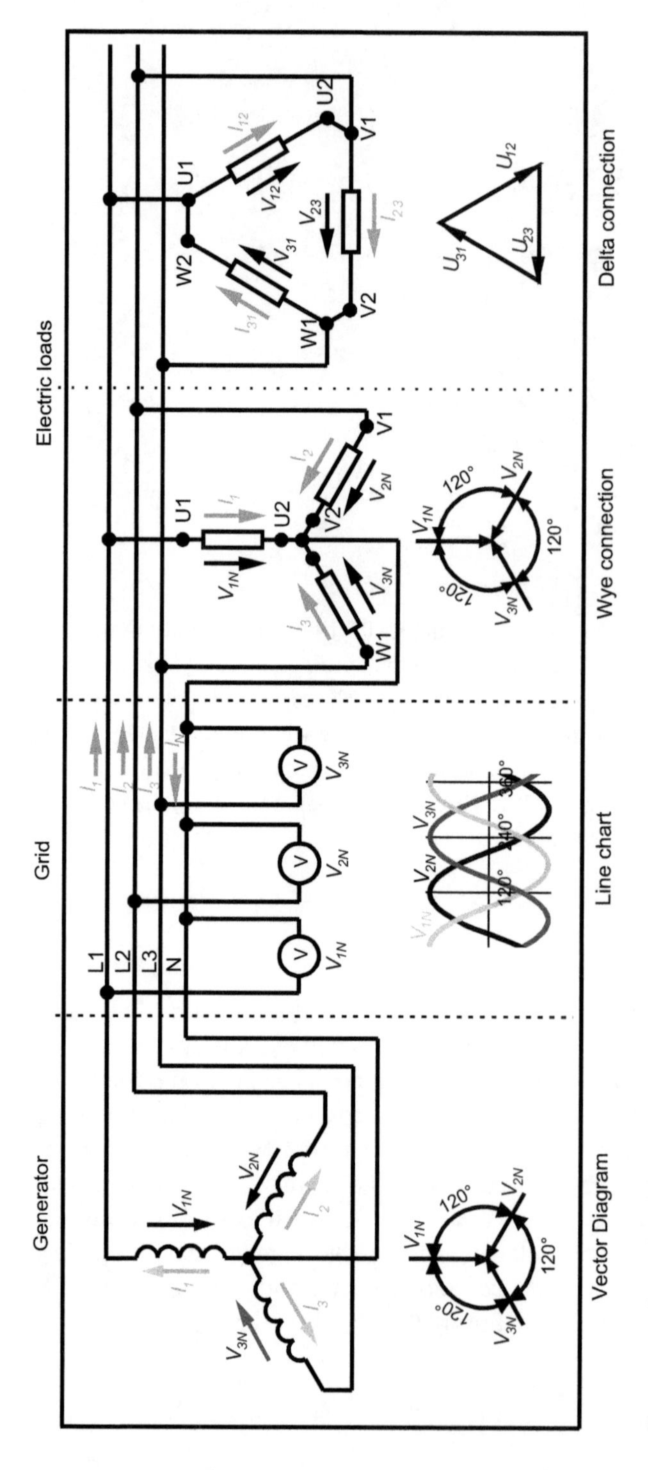

Figure 7.39: Three-phase electric system with line chart and vector diagram.

Derivation of the interlinking factor by means of a vector diagram

In Figure 7.40, the voltages V_{1N}, V_{3N} and V_{31} form an equilateral triangle with the base angle 30°. This triangle can be divided into two right-angled triangles. Using the trigonometric functions we derive:

$$\frac{V_{31}}{2} = V_{1N} \cdot \cos 30^\circ = V_{1N} \cdot \frac{\sqrt{3}}{2} \Rightarrow V_{31} = V_{1N} \cdot \sqrt{3}$$

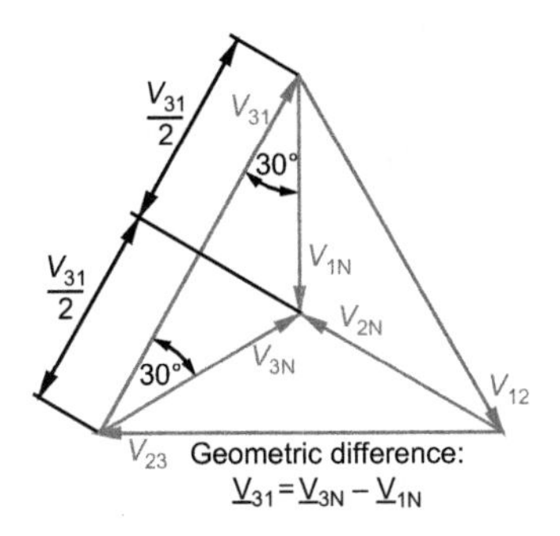

Figure 7.40: Vector diagram of voltages in a wye connection.

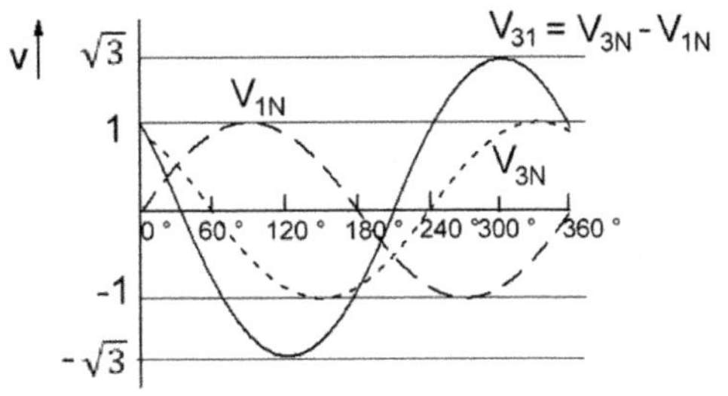

Figure 7.41: Phase voltage and conductor voltage in the curve diagram.

With respect to the vector addition in Figure 7.40, this results in:

$$\underline{V_{31}} + \underline{V_{1N}} - \underline{V_{3N}} = 0 \Rightarrow \underline{V_{31}} = \underline{V_{3N}} - \underline{V_{1N}}$$

If we form the difference of instantaneous values of the voltages v_{3N} and v_{1N} in the curve diagram of Figure 7.41, we obtain the development of the conductor voltages v_{31}. This is another example of how the peak value of the voltage v_{31} is greater by the factor $\sqrt{3}$ than the peak value of the voltage v_{1N}.

In a 400 V three-phase supply with 4 conductors the external conductor voltage is **V = 400 V**, the voltage between exterior conductor and neutral conductor is (V_{1N}, V_{2N}, V_{3N}) **230 V**. This enables the operation of three-phase loads with a higher power consumption and a rated voltage of 400 V (e.g. engines, electric stoves) as well as single-phase alternating current loads with a low power consumption of 230 V (e.g. electric bulbs or TVs) on the same grid.

7.8.2 Power of three-phase current

The power of a load when connecting three-phase currents can be calculated using the individual power of the three strings. Each of the three strings of the load is connected to the respective string voltage V_{Str} in the wye connection as well as in the delta connection and carries the string current I_{Str}. Therefore, the overall complex power with symmetrical load is: $S = 3 \cdot V_{Str} \cdot I_{Str}$

As the currents of the exterior conductors can be measured much more easily than the string currents, the values for conductor current and conductor voltage are inserted into the formula:

$$S = 3 \cdot V_{Str} \cdot I_{Str}$$

Table 7.3: Comparison of complex powers in wye and delta connection.

Three-phase current power under symmetrical load	
Wye connection	**Delta connection**
$I_{Str} = I \quad V_{Str} = \frac{V}{\sqrt{3}}$	$V_{Str} = V \quad I_{Str} = \frac{1}{\sqrt{3}}$
$S = 3 \cdot I_{Str} \cdot V_{Str} = 3 \cdot I \cdot \frac{V}{\sqrt{3}}$	$S = 3 \cdot I_{Str} \cdot V_{Str} = 3 \cdot V \cdot \frac{I}{\sqrt{3}}$
$S = \sqrt{3} \cdot V \cdot I$	$S = \sqrt{3} \cdot V \cdot I$

The powers of the wye and the delta connection are calculated using the same formulae. For complex power S, active power P and reactive power Q we receive:

$$S = \sqrt{3} \cdot V \cdot I$$

$$\boldsymbol{P = \sqrt{3} \cdot V \cdot I \cdot \cos \varphi}$$

$$Q = \sqrt{3} \cdot V \cdot I \cdot \sin \varphi$$

S Complex power in VA
V Conductor voltage in V
I Conductor current in A
P Active power in W
Q Reactive power in var
$\cos \varphi$ Active power factor
$\sin \varphi$ Reactive power factor
φ Phase shift angle in °

? 7.9 Review questions

1) How do you calculate the inductive reactance of a coil and how do you calculate the inductive reactance of a capacitance?

a) What regularity applies for the phase shift between current and voltage, respectively?
2) What fundamental calculation rules are there?
   a) for the series connection of R, X_L and X_C ?
   b) for the parallel connection of R, X_L and X_C ?
3) What is the effective value?
   a) Why do we state the effective value of an alternating quantity?
   b) What is the mathematical correlation between the effective value and the peak value of an alternating quantity?
4) Name the performance terms in the alternating-current circuit.
   a) What is the correlation between the performance terms?
   b) How do you calculate the active power? What does the power factor reveal?
5) What is a three-phase current?
   a) What is the correlation that is stated by the interlinking factor?

7.10 Exercises

EXERCISE 7.1

Calculate from an alternating voltage (sinus-wave) with a frequency f of 50 Hz and an effective voltage V_{RMS} of 24 V the peak value V_p, crest factor F_C and duration of one period T.

EXERCISE 7.2

When the active power of a load is 400 W and the reactive power is 100 Var, which apparent (complex) power has to be delivered from the grid?

EXERCISE 7.3

Calculate the impedances X_C and X_L of a capacitance C = 0.47 µF and an inductivity L = 50 mH at 50 Hz and at 2 kHz.

EXERCISE 7.4

Calculate the total impedance X_{tot} when above X_C and X_L were connected in parallel. How big were the currents I_C and I_L when a voltage source of 12 V is connected? How big is the total current I? At which frequency f are the impedances equal and how big is the total current in this resonance case?

EXERCISE 7.5

A series connection of an ohmic resistor R = 100 Ohm and an inductivity L is given. When a current I of 1 A and a total voltage V of 230 V at a frequency f of 50 Hz is measured how big is the inductivity, which voltages can be measured at the resistor U_R and at the inductivity U_L? Build a vector diagram.

Exercise 7.6

A series connection of an ohmic resistor R = 3300 Ohm and a capacitance X_C = 5000 Ohm is given. When a voltage U_R of 6 V and a frequency f of 1 kHz is measured how big is the current I and the total voltage U_{tot}, which voltage can be measured at the capacitor U_C? Build a vector diagram.

Exercise 7.7

A parallel connection of an ohmic resistor R and an inductivity L = 50 mH is given. When a voltage V of 24 V and the currents at the resistor I_R = 25 mA and at the inductivity I_L = 10 mA is measured how big is the resistor, which total current I_{tot} can be measured and which frequency f is present? Build a vector diagram.

Exercise 7.8

A parallel connection of an ohmic resistor R and a capacitance C is given. When a voltage U of 24 V, a total current I_{tot} = 1 A and a frequency f of 400 Hz as well as a phase angle of 35° is measured how big are the currents I_C and I_L and which values have the resistor and the capacitance? Build a vector diagram.

8 Fundamentals of measurement and regulation technology

Components in all areas of technology need to have specific properties in order to be suitable for their intended use. To determine physical quantities such as length, voltage, current, resistance, mass, force, temperature, pressure, time etc. and their uncertainty, we need the devices and methods **provided, described and standardised by** measurement technology.

With the help of **regulation technology**, we can influence a system in a way that a specific quantity follows a (variable) desired set-point value, independent of disturbances.

! The following fundamental terms need to be distinguished in measurement technology:

- **Measurement**: Derive a quantitative statement about a physical quantity.
- **Counting**: Determine the number of equivalent events over a specific time period.
- **Controlling**: Detect if a device complies with the specified properties and/or if these are within the specified margin of error.
- **Calibration**: Determine and document the deviation of a measured value from the true value.
- **Adjustment**: Set the measuring devices to a specific value.
- **Calibration verification**: Officially set margins of error must be complied with. The calibration verification is the process of eliminating the deviations determined during calibration (adjustment to the "true value", which is determined using a measuring device with a higher level of accuracy). The calibration verification is performed by an approved testing laboratory and has to be repeated on a regular basis.
- The **measuring range** of a measuring device, e.g. 750 V, is the area of measured values where the set margins of error, e.g. ±2 %, are not exceeded. When determining an unknown measured value, we, at first, choose the widest measuring range possible (for security reasons) of the quantity that we want to measure, e.g. voltage. We then adjust to the measuring range where the meter is in the top third of the scale (for meters with an analogue display). Measuring devices are divided into analogue and digital display meters.
- The **accuracy class** refers to the expected maximum deviation (percentage of the nominal full scale) of the measured value from the true value of the physical quantity to be measured.
  Example: A device with the accuracy class 2.5 has a maximum deviation of 2.5% from the nominal full scale.

https://doi.org/10.1515/9783110521115-008

There are **analogue measuring devices** (measurement display by means of a needle) and **digital measuring devices** (measured values can be directly read off a display). Both types are justified in their respective areas of application. Analogue meters are well suited to measure time-varying measured values (e.g. speedometer: speed of a car). For static measured values, digital measuring devices are more convenient as no reading errors[94] are possible. Many digital measuring devices are equipped with an automatic measuring range switchover and the advantage of analogue measuring devices (good representation of time- varying quantities) is emulated by representing it with a bar on the display.

The following sections introduce a few important concepts and measuring methods for certain quantities. However, they only provide a limited insight into the vast field of measurement technology.

8.1 Measuring electrical quantities

Electrical quantities (voltage, current and resistance) are easily measured using electrical measuring devices, as long as the quantities remain within the measuring ranges. For example, voltage in the mV range can still be measured well as they can easily be proportionally increased using amplifier circuits. Voltages in the µV range, however, can only be measured with great effort as disturbances (caused by induction, contact resistances, line resistances) have a negative impact. Very high voltages and currents (in the kV and kA range) can be made measurable using voltage or current transformers. Measuring such voltages or currents directly would destroy the device.

Measurement transformers

Measurement transformers are low power transformers, which convert an input quantity into an output quantity according to a pre-set relation. They transform high voltages or currents into measurable values and ensure galvanic isolation.[95]

- **Voltage transformer**: e.g. high voltage 10 kV to measuring voltage 100 V
- **Current transformers:** used to measure high currents in low-voltage systems or currents in high-voltage systems for security reasons
- **Current clamp transformer**: The clamp is the iron core of the transformer; the conductor corresponds to the input winding of the current transformer

---

94 e.g. Parallax error: It forms when you do not look at the needle of an analogue measuring device vertically with the scale behind. Resolve to recognize and prevent this mistake: Mirror behind the needle.

95 This leads to an increase in accuracy for the measurer.

8.1.1 Electrical meters

For an analogue measuring device to represent a measured value, a meter with a needle and a scale is used. It generates the torque required for the needle deflection in the moving component. Meters differ in their structure and in their properties. To gain insight into the functioning of meters, Table 8.1 illustrates three fundamental types of meters.

Table 8.1: Fundamental types of meters.

Meter	Structure/function	Properties
Moving-coil ammeter Symbol: 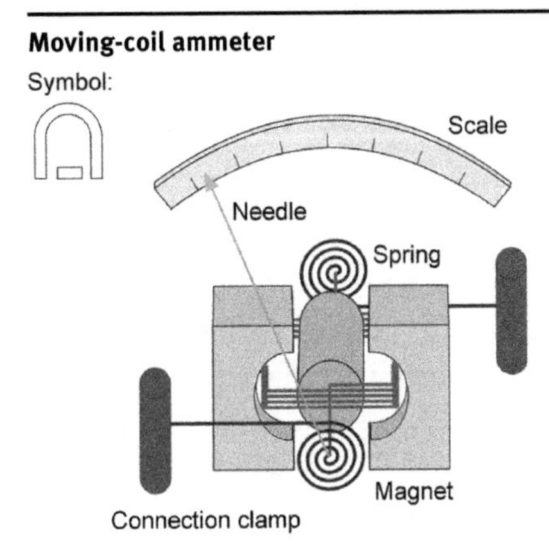 **Figure 8.1:** Moving-coil ammeter.	**Structure/function:** In the magnetic field of a permanent magnet, a coil is rotatably mounted (moving coil). The current flowing through the moving coil (caused by the voltage to be measured) generates a magnetic field that has an angle to the magnetic field of the permanent magnet. The interaction of these two magnetic fields creates a torque that deflects the needle attached to the coil until the deflection equals the torque of the spring working in the opposite direction.	**Properties:** Displays the **arithmetic mean** of voltage (or current) due to the inertia of the meter. High accuracy and sensitivity Linear scale Low self-consumption
Moving-iron ammeter Symbol: **Figure 8.2:** Moving-iron ammeter.	**Structure/function:** Within the coil, there is a stationary iron core and a moving iron core to which the needle is attached. If a current pass through the coil, both cores are magnetised in the same direction, hence, they repel each other. The moving iron core shifts away from the stationary iron core (needle deflection) until the restraining moment of the spring equals the torque caused by the repulsion.	**Properties:** Measures the **effective value (RMS)** of a voltage or current Non-linear scale division Only up to a frequency of approx. 400 Hz.

Table 8.1 (continued)

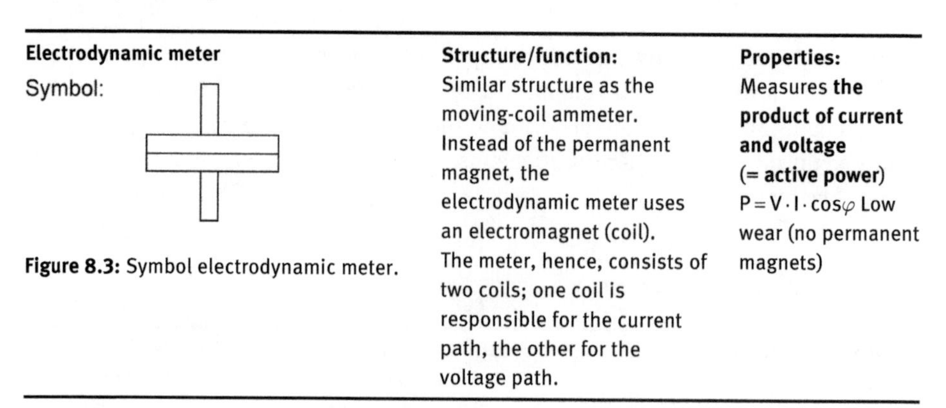

Electrodynamic meter	**Structure/function:**	**Properties:**
Symbol: **Figure 8.3:** Symbol electrodynamic meter.	Similar structure as the moving-coil ammeter. Instead of the permanent magnet, the electrodynamic meter uses an electromagnet (coil). The meter, hence, consists of two coils; one coil is responsible for the current path, the other for the voltage path.	Measures **the product of current and voltage** (= **active power**) $P = V \cdot I \cdot \cos\varphi$ Low wear (no permanent magnets)

8.1.2 Digital meters

The digital measuring of voltages is principally based on the counting of impulses (see Figure 8.4). The voltage to be measured V_x and a defined sawtooth voltage are applied to a comparator. With the beginning of the start-up of the sawtooth voltage, a gate circuit opens and allows the impulses of a generator to pass through as long as V_x is lower than the sawtooth voltage. The impulses are registered by a counting meter and, after decoding, they are represented on a display. As soon as the sawtooth voltage reaches the value V_x, the gate shuts. In case the voltage V_x is low, the sawtooth voltage

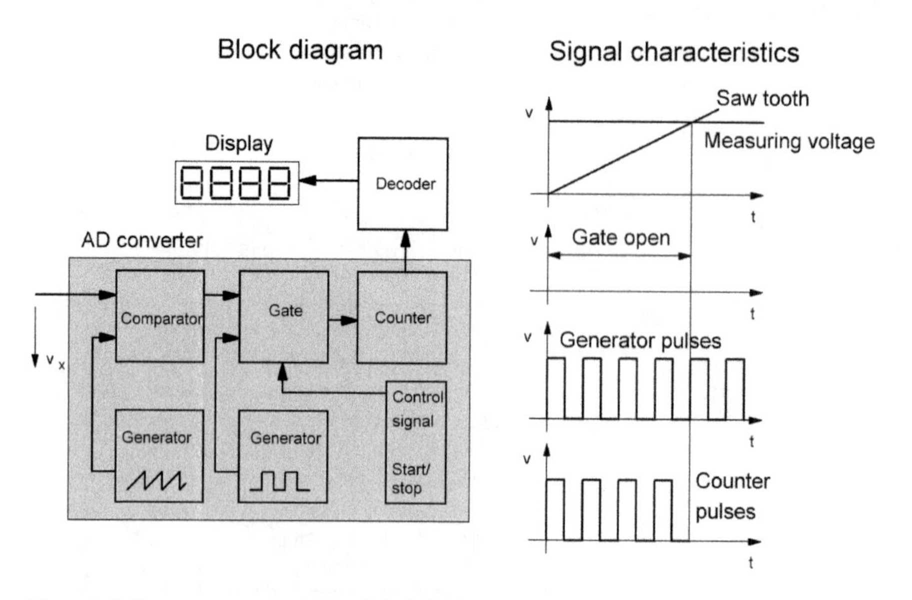

Figure 8.4: The core component of a digital display meter is the analogue-to-digital converter (ADC, A/D, or A-to-D).

has reached this value after a short period of time; however, if V_x is high, many impulses can pass through the gate. The number of counted impulses, hence, is a measure for the voltage V_x we want to determine.

8.2 Measuring non-electrical quantities

Non-electrical quantities can be measured using the properties of their materials or known physical phenomena that lead to a change in or the generation of an electrical quantity (e.g. resistance, voltage). Due to the known (or determinable) correlation between non-electrical quantity and electrical measured quantity, also non-electrical quantities can be measured using electrical measuring devices.

8.2.1 Length measurement, fluid-level measurement

For length measurement we use ohmic, capacitive or inductive measuring transducers.

Resistive sensor (length)

At a resistive element, the contact tap is moved along the distance s. The resistance between terminal 1 and terminal 2 is proportional to the distance s.

The following applies:

$$R_S = \frac{R_{tot}}{l} \cdot s \Rightarrow s = \frac{R_S}{R_{tot}} \cdot l$$

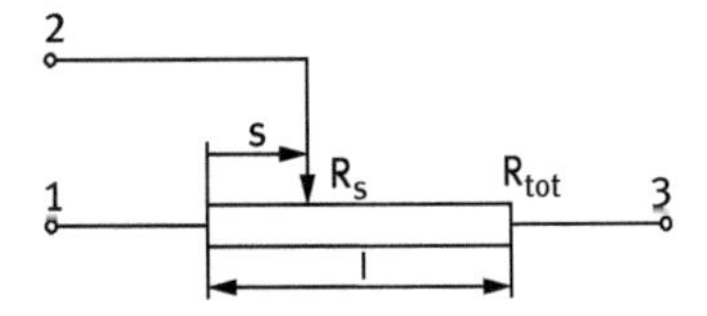

Figure 8.5: Resistive sensor – general structure.

Inductive sensor (length)

By moving the iron yoke, the magnetic resistance changes and, consequently, so does the inductance. If a sinusoidal alternating voltage with a constant frequency is applied to the coil, this results in:

The effective value of the measured current corresponds to a distance s. Due to the non-linearity of $L(s)$, the length measurement is only reasonable for short Δs with a stationary operating point s_0.

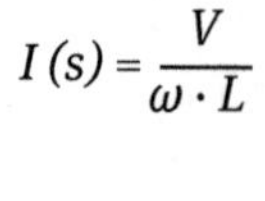

$$I(s) = \frac{V}{\omega \cdot L}$$

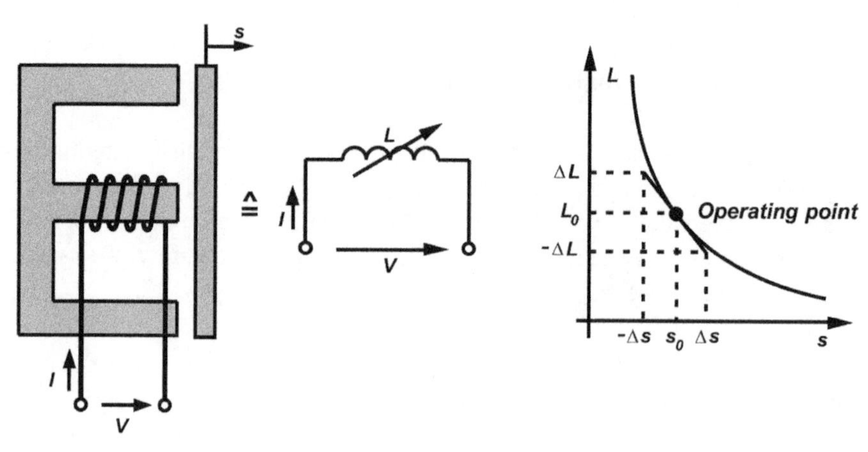

Figure 8.6: General structure of an inductive sensor used to measure length.

Capacitive length measurement

Analogue to the inductive sensor for length, we modulate a voltage with a known constant frequency and measure the current that is dependent on the capacitive resistance $X_C(s) = \frac{1}{\omega \cdot C(s)}$:

$$I(s) = \frac{V}{X_C} = V \,.\, \omega \,.\, C(s)$$

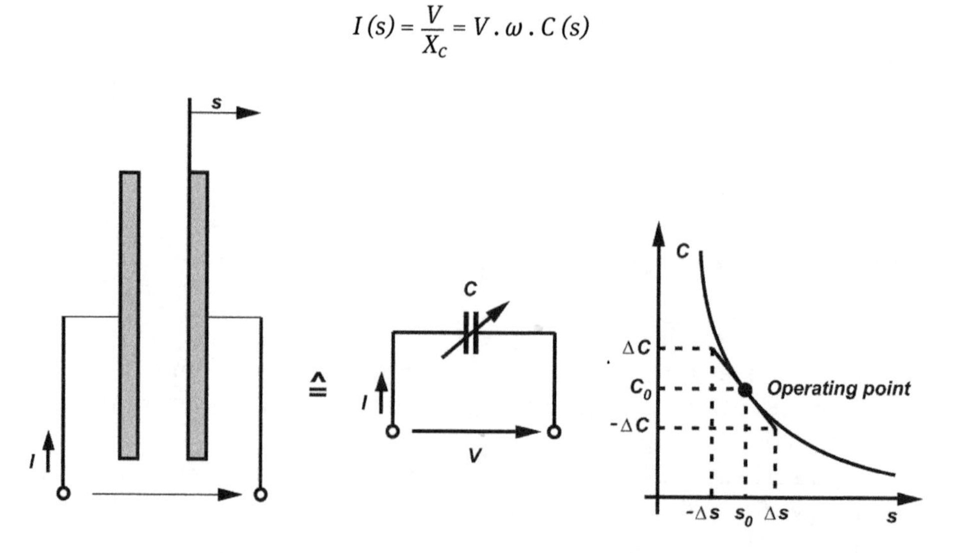

Figure 8.7: General structure of a capacitive sensor to measure length.

8.2.2 Fluid-level measurement

Since the capacitance of a plate capacitor depends in the dielectric between its plates, it can be used to determine the fluid level of a container.

Two parallel metal plates (equivalent to a plate capacitor) with the length l are submerged into a liquid. Depending on the depth of submersion, the capacitance of the capacitor changes as it can be regarded as parallel connection of two capacitors. The plate length of C_1 is $l - l_1$ and its dielectric is air ($\varepsilon_{air} \approx \varepsilon_0$). The plate length of C_2 is l_1 and the dielectric constant of the liquid is $\varepsilon_{liquid} = \varepsilon_0 \cdot \varepsilon_{r,\ liquid}$.

Therefore, the total capacitance of the plates depends on the depth of submersion.

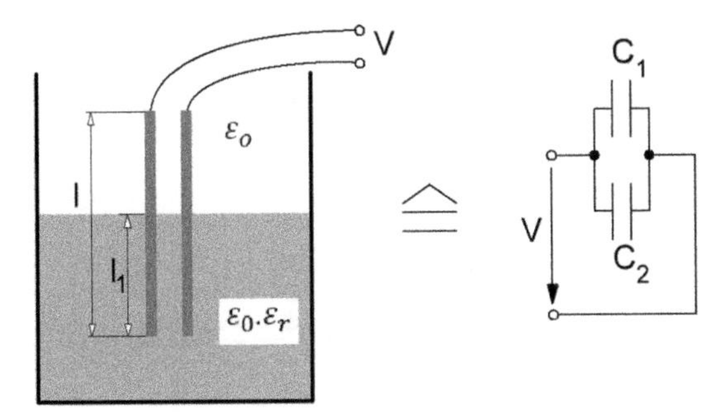

Figure 8.8: Principle of fluid-level measurement due to capacitance change.

8.2.3 Force measurement

The force is measured using strain gauges (=SG).

Strain gauges (SG)

Strain gauges (SG) are not only important for measuring strain but they are also used to measure all kinds of mechanical measured quantities that are related to the strain of elastic spring bodies such as forces, pressures, distances, accelerations, bending moments, torques ...

SGs consist of metal resistance wires with a diameter of approx. 20 μm, metallic resistance foils or semiconductor materials. Their measuring principle is based on the great alteration in resistance when they are deformed.

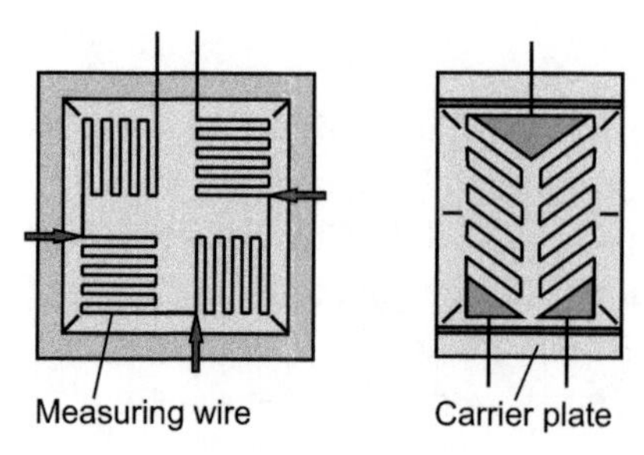

Figure 8.9: Different designs of strain gauges.

8.2.4 Velocity and rotational speed

Generally, the momentary speed v is given by:

$$v = \frac{ds}{dt}$$

Speedometer generator

Rotational speed of several s^{-1} up to approx. 10,000 min^{-1} are usually measured using speedometer generators. They are either direct-voltage or alternating-voltage generators.

The measured voltage V is proportional to the rotational speed n according to the law of induction. Since $v = \omega \cdot r$, the circumferential velocity v of a shaft is therefore also proportional to the rotational speed n.

$$V = K \cdot n$$

$$v = K_1 \cdot n$$

V Voltage in V
K Proportionality constant
n Rotational speed in s^{-1}
v Velocity in m/s
K_1 Proportionality constant

The rotational speed or velocity can be determined using toothed lock washers with inductive sensors, toothed lock washers with photoelectric sensors or with magnetic discs with a Hall effect sensor.

8.2.5 Temperature

Resistance thermometer

The resistor of a resistance thermometer (nickel or platinum wire, wound around mica or phenolic paper plates) changes with temperature.
Determining the resistance and consequently the temperature (known correlation) happens using current and voltage measurement or a measurement bridge.

For a Pt100 element with $\vartheta = 0°C$, the following applies:

$$R_0 = 100°C$$

For $0°C < \vartheta < 100°C$ applies:

$$R_\vartheta = R_0 \cdot (1 + \alpha_0 \cdot \vartheta)$$

For $\vartheta > 100°C$ or smaller errors:

$$R_\vartheta = R_0 \cdot \left(1 + \alpha_0 \cdot \vartheta + \beta_0 \cdot \vartheta^2 + \gamma_0 \cdot (\vartheta - 100°C)\right) \cdot \vartheta^3$$

with the coefficients:

$$\alpha_0 = 3.9083 \cdot 10^{-3} \cdot K^{-1}$$

$$\beta_0 = -5.775 \cdot 10^{-7} \cdot K^{-2}$$

$$\gamma_0 = -4.183 \cdot 10^{-12} \cdot K^{-4}$$

Thermocouple

In contrast to resistance thermometers,[96] thermocouples are active sensors, which means that they generate a voltage. Said voltage is referred to as thermoelectric voltage.

Physical background:

When two dissimilar metals contact each other, electrons diffuse from one metal into the other with varying depth. The metal which emits more electrons to the other metal than it gains is charged positively compared to the other metal. The resulting contact voltage is called "Galvani potential". The difference in the diffusion of electrons grows with increasing temperature. This is called Seebeck effect or thermoelectric effect.[97]

---

96 Resistance thermometers are passive sensors as the resistance (which is proportional to the measured quantity temperature) can only be determined through applying a constant voltage and measuring the current (or vice versa). You figuratively have to "ask" for the response of the current through the voltage – they "do not talk actively".

97 The reverse effect to thermoelectric effect is the Peltier effect, during which a voltage is applied and thus heat is "pumped" from one welding spot to the other. Such a structure ("Peltier element") works as a heat pump and is used for cooling purposes (e.g. in cool boxes).

Thermocouples consist of two wires made of dissimilar types of metals or alloys. The endings of the wires are welded together at a so-called welding spot. A thermoelectric voltage is generated (which is the difference between the two contact voltages) that can be measured using a high-resistance[98] voltmeter or a compensator.

Thermocouples are very small, almost punctiform (welding spot) measuring points for temperature of up to 1,000 °C – 1,500 °C depending on the material.

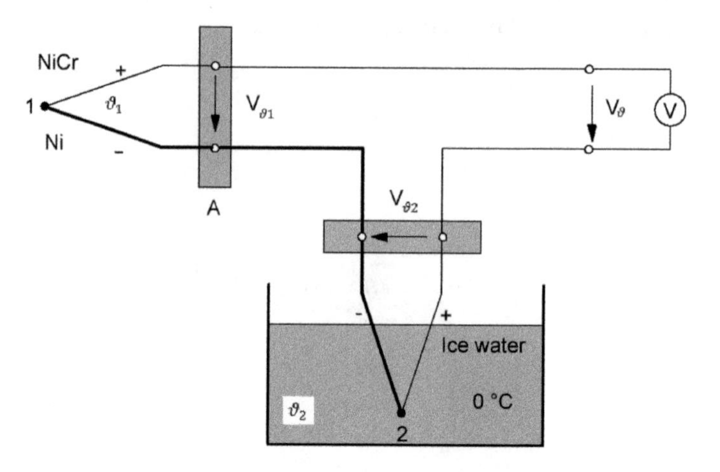

Figure 8.10: Measuring arrangement with ice water (triple point H_2O = 0 °C).

The measurable thermoelectric voltage corresponds to **the temperature difference** between measuring point 1 and measuring point 2. The temperature of measuring point 2 is kept at a known reference temperature or the temperature of measuring point 2 is measured with another procedure (e.g. a resistance thermometer Pt100) and the measured result is corrected.

NiCr/Ni elements (= "type K") e.g. have a thermoelectric voltage of 43.31 mV with a 1,000 °C temperature difference between measuring points 1 and 2.

A thermocouple has the advantage over a resistance thermometer that the volume to be heated at the measuring point is very small and thus fast changes in temperature are tracked with almost no delay in the measurement value output. Furthermore, very high temperatures (up to 2,500 °C) can be measured, depending on the type of the thermocouple as resistance thermometers can only withstand a relatively low temperature (up to 600 °C) without being destroyed.

98 "High-resistance" means that the measuring device has a high internal resistance. Thus, almost no current flows and no voltage drops at the internal resistance of the voltage source to be measured, which would distort the measurement.

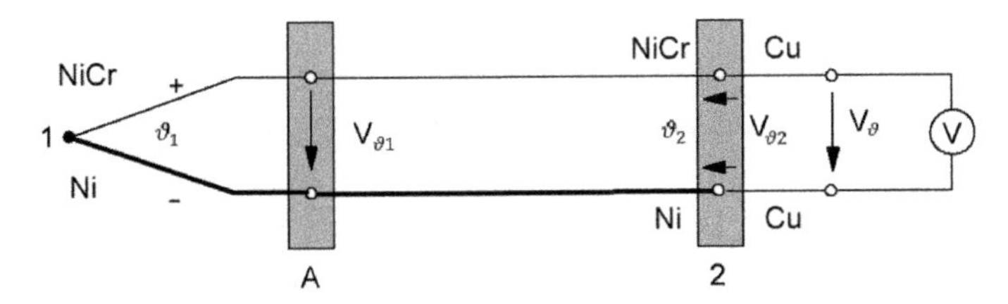

Figure 8.11: Thermocouple with known reference junction ("cold junction compensation").

8.2.6 Photometry

Photoresistor

Different semiconductors with intercalated foreign atoms increase their conductivity with exposure to light proportional to the illuminance, e.g. cadmium sulphide (CdS), cadmium selenide (CdSe), indium antimonide (InSb).

The more light falls onto the photoresistor, the lower is its electrical resistance. The cause for this function is the inner photoelectric effect in a layer consisting of amorphous semiconductors. Compared to other photosensors, photoresistors react very slowly.

Figure 8.12: Photoresistor made of cadmium sulphide.

Photoelectric cell

Discharge tubes that are evacuated or filled with noble gas (little pressure) count among the passive photoelectric cells. When exposed to light, electrons are released through photoemission and the electrons migrate towards the anode. The cathode consists of e.g. sodium, potassium, caesium, . . .

Photodiode

During light absorption in a semiconductor depletion layer, electrons become mobile due to the depletion-layer (photoelectric) effect and migrate through the depletion layer. A photoelectric voltage is generated.

Application: Illuminometer (photometer)

8.3 Bridge circuit

A bridge circuit[99] consists of a parallel connection of two voltage dividers R_1, R_2, R_3 and R_4 (see Figure 8.13). Both voltage dividers are connected to a joint voltage source. The distance between the tapping points A and B is called bridge branch or diagonal bridge. Between the two points A and B a bridge voltage occurs.

Bridge circuits are used in measurement technology to determine an unknown resistance R_x (e.g. temperature dependent resistance, strain gauge) using three known resistances. The unknown resistance can only be determined if the bridge is balanced: the potentials of the nods A and B were equal and no current flows through the bridge branch.

In a slide-wire bridge (see Figure 8.14), the sliding contact is adjusted until the bridge is balanced. The position of the tap allows us to read off the resistance value R_X.

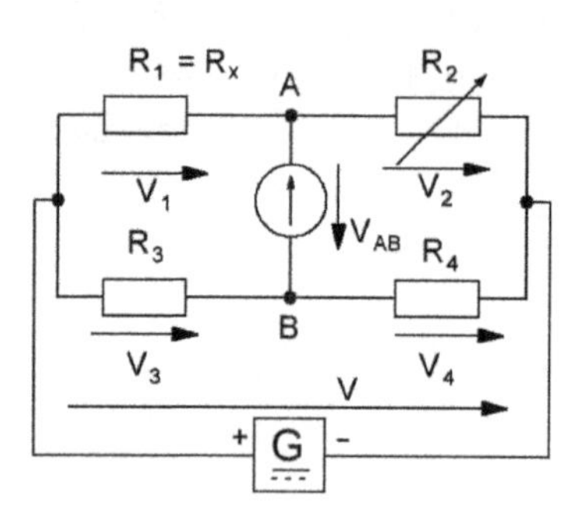

Figure 8.13: Bridge circuit.

$$R_X = R_2 \cdot \frac{R_3}{R_4}$$

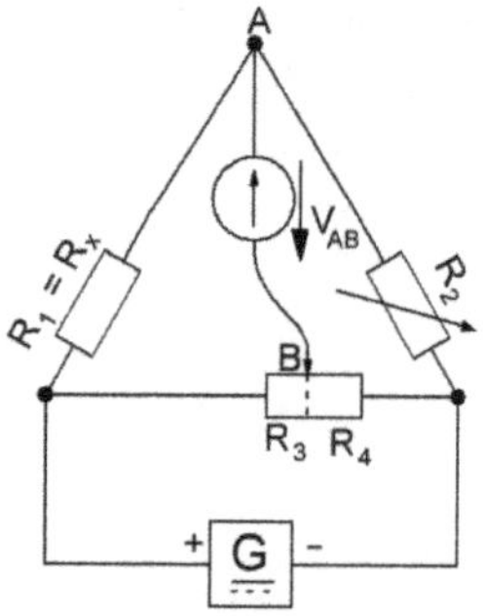

Figure 8.14: Slide-wire measurement bridge.

A bridge circuit is balanced when there is no current flow in the bridge diagonal, i.e. when the resistance ratio of the two voltage dividers is equal:

$$V_{AB} = 0, if \ \frac{V_1}{V_2} = \frac{V_3}{V_4} \ or \ \frac{R_1}{R_2} = \frac{R_3}{R_4}$$

R_1, R_X Unknown resistance
R_2 Comparative resistance (adjustable)
R_3, R_4 Bridge resistors

99 Variants are H circuit, H bridge, Wheatstone bridge or full bridge.

8.4 Control engineering

8.4.1 Regulation technology vs. control technology

Regulation technology, besides control technology, is a sub-discipline of automation technology. In contrast to control technology, in regulation technology there is a feedback of the output quantity to the input of a regulator (see Figure 8.15 and Table 8.2). Therefore, disturbances ideally do not affect the output quantity as the control variable is adjusted accordingly. This is why a controller is not able to react to external disturbances. Regulation or regulating is a process during which the quantity to be regulated (regulating variable) is continuously measured, compared to another quantity (reference variable = set-point value) and – depending on the result of the comparison – manipulated to correspond to the reference variable for the sake of balance. The resulting action flow happens in a closed loop (control loop).

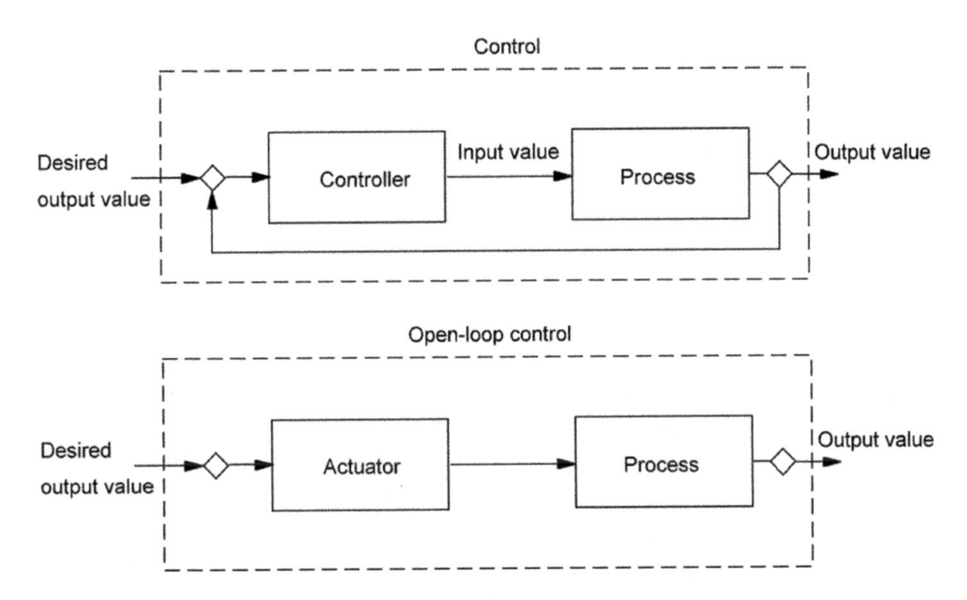

Figure 8.15: Comparison regulation/control technology.

Table 8.2: Comparison regulation/control technology.

Examples of control technology	Examples of regulation technology
A microwave is set to 2 minutes to defrost frozen food. Consistency and freeze condition of the frozen food are not considered.	Air conditioning: A chosen temperature set-point value is retained. Depending on the temperature, a cooling/heating element is turned on or off.
A toy robot is programmed to run into one direction. It behaves according to its programming in spite of obstacles blocking its way.	Steering a car: Oversteer or understeer is corrected (= feedback) by the driver adapting the steering angle.
The lawn irrigation system is turned on automatically at a set time, even if it is raining at that moment.	The lawn irrigation system is turned on/off according to the moisture of the lawn.
Washing machine: The programme system (in non-electronic washing machines) switches the washing drum in right-hand or left-hand rotation. Spinning activates the drain pump, turns the heating element off/on and turns the fresh-water valve on/off – irrespective of whether the switched components are functioning.	Centrifugal governor of a steam engine: The steam supply valve is opened or closed according to the rotational speed of a shaft with a centrifugal governor attached to it.

8.4.2 Standard control loop and terms

Regulation is tasked with bringing a physical quantity, e.g. temperature, pressure, filling level or voltage to a given value and holding that value even if there are disturbances.

In Figure 8.17, the voltage of a direct current generator is regulated manually. If the voltage V declines due to an increase in load current I, the excitation current must be increased. If the voltage increases, the excitation current I_e must be reduced. In terms of regulation technology, voltage is the regulating variable x, the excitation current is the control variable y and the load current I is the disturbance variable z.

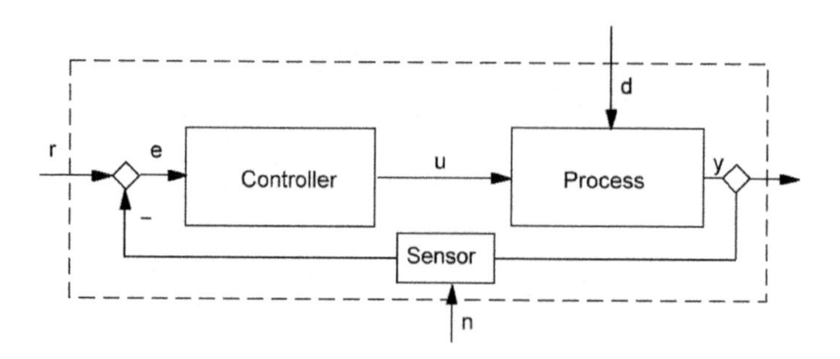

Figure 8.16: Standard control loop.

The reference variable w (set-point value) is marked on the voltmeter. To hold the voltage at constant, the operator monitors the voltmeter and compares the measured value to the marked set-point value. In the case of a deviation (system deviation e), the operator will adjust the excitation current accordingly. The greater the system deviation, the greater the change of the control variable y.

If the operator is substituted by a device that can autonomously hold the regulating variable at its set-point value, we talk about an automatic regulating device.

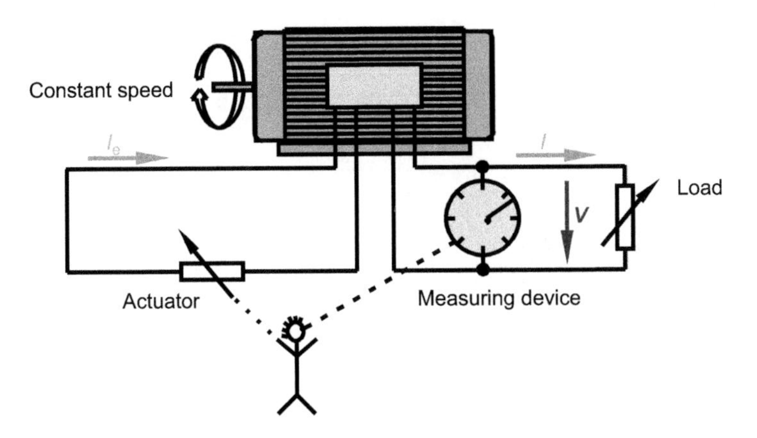

Figure 8.17: Manual regulation of the generator voltage.

Important technical terms

Controlled system: The part of the control loop where the regulating variable x is regulated.
Regulator: It compares the regulating variable x and the reference variable w to each other and forms the control variable y.
Actuator: It serves to set the control variable
Regulating device: The entirety of regulator, actuator and measuring device.
A regulation is a closed sphere of action (control loop). The regulating variable x is measured continuously and compared to the reference variable w. The result of the comparison, the system deviation e, is processed by the regulator and returned to the input of the control path as modified control variable y.

We differentiate between:

Continuous regulators
- P regulator
- I regulator
- D regulator
- PD regulator
- PID regulator

Discontinuous regulators
- Two-point regulators
- Three-point regulators

8.4.3 Controller (I, D and PID)

P controller

By means of fluid-level control (see Figure 8.18), the behaviour of a P controller is demonstrated. If the fluid level increases, the float rises and closes the valve by lifting the lever. At the fluid level x_e the valve is completely closed, at x_a the valve is completely open. In the range between x_a and x_e a change in the valve position Δy is proportional to the change in fluid level Δx.

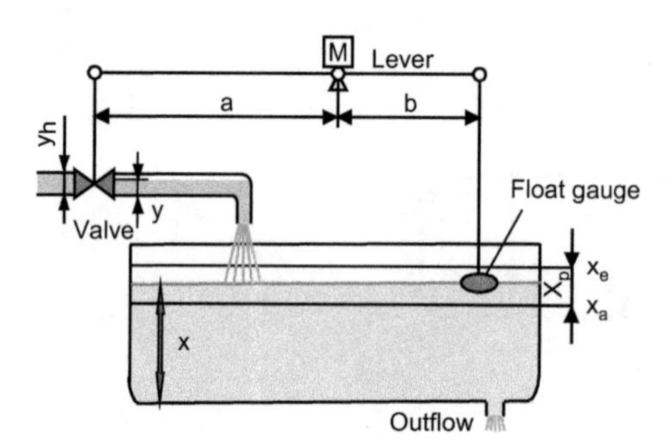

Figure 8.18: Fluid-level control using a P regulator.

When shifting the pivot point M of the lever to the left, the fluid level must be increased yet again to fully close the valve and be dropped ever more to fully open it. The range between x_a and x_e is extended. By shifting the pivot point to the right, the range between x_a and x_e is narrowed. The difference between x_a and x_e is called **proportional range X_p.**

The P regulator is fast, but it has a remaining system deviation. The greater K_{RP}, the smaller the remaining system deviation e_p.

P controller

- creates a **control variable** that is proportional to the control deviation
- reacts fast and never settles fully to equilibrium → **permanent control deviation**
- is very simple and inexpensive.

I controller

I controller (integral controller)

- creates a control variable that is proportional to the temporal integral of the control deviation; the change in the control variable at constant control deviation is a ramp.
- reacts slowly. It settles abruptly changing reference and disturbance variables to equilibrium → no permanent control deviation.
- has a tendency to overshoot; the stability behaviour declines

---

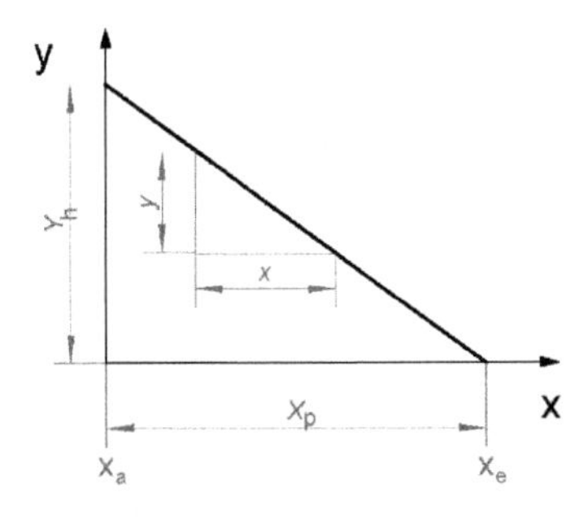

block diagram: e → [ ] → y

$$K_{RP} = \frac{\Delta y}{\Delta x} = \frac{Y_h}{X_p}$$

K_{RP} proportional coefficient of the controller
Δy change in the regulating value
Δx change in the control value
Y_x adjusting range
X_P proportional range

Figure 8.19: Characteristic curve, proportional coefficient and block diagram of a P controller.

D controller (differential controller)

Operating principle: In case the float position (system deviation e) changes abruptly at the fluid-level control (see Figure 8.20), the oil pressure shock absorber is deflected. The spring pulls back the shock absorber and the valve into the initial position, even when the system deviation is still applied.

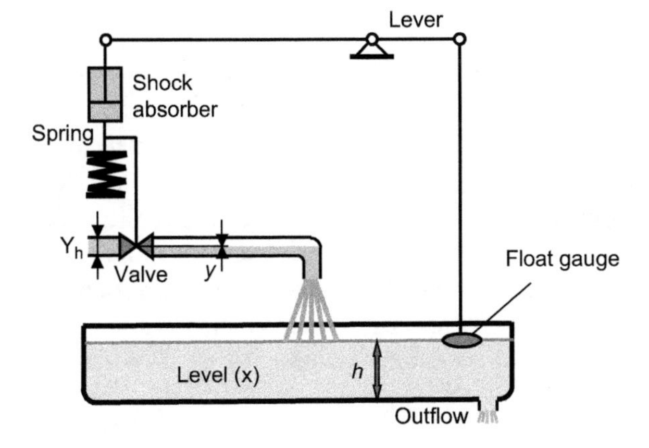

Figure 8.20: Fluid-level control using a D controller.

Dynamic behaviour: During a step response, the control variable immediately reaches the maximum value and instantly drops back to zero if the system deviation remains the same.

D controller
- generates a control deviation that is proportional to the **temporal change** of the control deviation; the control variable therefore disappears with constant control deviation.
- reacts quickly → stability behaviour↑
- is only used in conjunction with other regulators as it does not settle to equilibrium at all

PID controller

Dynamic behaviour:

The P part reacts immediately to a system deviation. The I part prevents a permanent system deviation and the D part reacts to the rate of change of the system deviation.

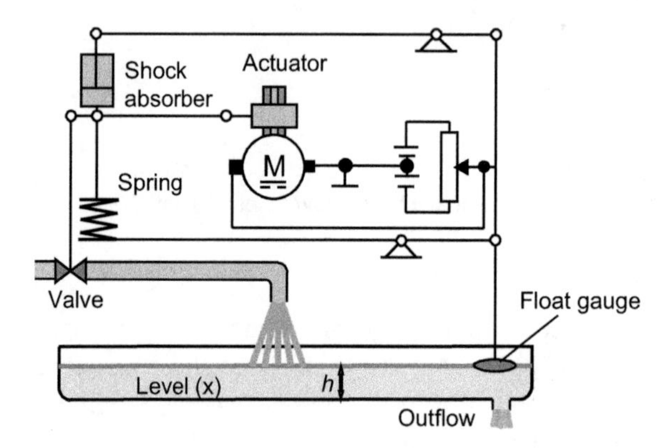

Figure 8.21: Fluid-level control using a PID controller.

The PID controller is the most universal of the classic regulators and combines the good properties of the other regulators. The PID regulated loop is accurate and very fast. Therefore, in most applications a PID controller is used. They are sold by manufacturers as finished digital devices. The set-point value and the regulator parameters (*Kp*, *Ki* and *Kd*) can[100] be adjusted.

With **operational amplifiers** (see chapter 0), analogue controllers (see Figure 8.23) can be set up. The regulatory function is implemented through the wiring of the operational amplifier with the corresponding components (resistors and capacitors). The control parameters are adjusted through selection of the resistance and capacitance values.

100 Usually many more parameters and functions can be adjusted.

It is also possible to realise a PID regulator with a software (e.g. Labview[101]). However, a connection with physical reality is necessary (interface card or hardware). Regulator equation:

$$y(t) = K_p \cdot e(t) \cdot + K_i \cdot \int_0^t e(\tau) \cdot d\tau \cdot + K_d \cdot \frac{de(t)}{dt}$$

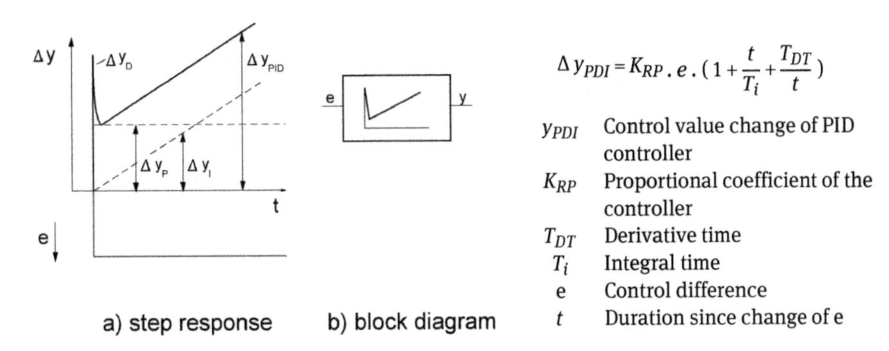

Figure 8.22: Step response, block diagram and formula for the control variable of a PID controller.

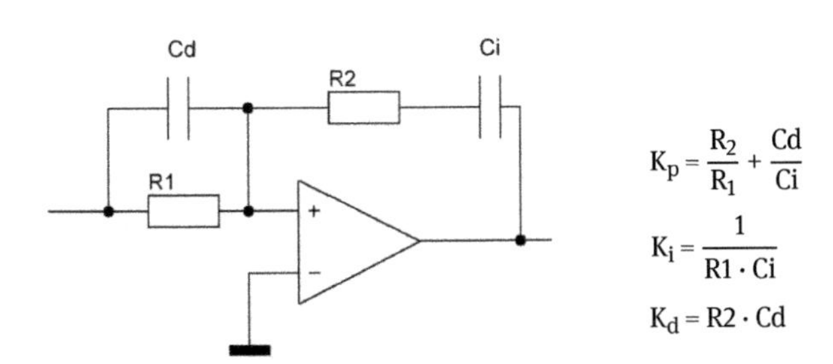

Figure 8.23: Analogue PID controller realised using an operational amplifier.

101 In Labview there is a so-called "virtual instrument" (VI) which provides the PID regulatory functions for implementations with software in the programme.

? 8.5 Review questions

1. What electrotechnical possibilities to measure temperature do you know?
2. How does a bridge circuit work and what can you measure with it?
3. Explain the operating principle of a wire strain gauge (WSG).
4. What is the fundamental difference between regulating and controlling?
   a) What does a simple control loop look like?
   b) What is a PID regulator?

9 Electric machines

Electric machines use the properties of electromagnetic interaction. They are based on electromagnetic induction and magnetic forces, which are described as Lorentz force and, in some types of machines, as reluctance force.

With **rotating electric machines**, magnetic forces play a vital role. Their purpose is to transform electric power into mechanical power at a shaft. If electric power is converted into mechanical power, we talk about an electric motor. If, vice versa, mechanical power is converted into electric power, we talk about an electric generator. Some types of electric machines can be operated not only as a motor, but also as a generator; the actual function is determined by the area of application of the machine.

With **resting electric machines,** magnetic forces only play a subordinate or undesired role, as no movements are carried out and the function of the transformer consists of transforming alternating voltages between different voltage levels or for galvanic isolation.

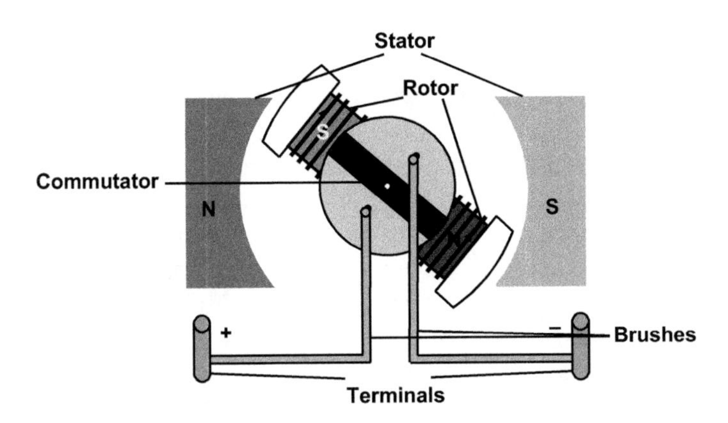

Figure 9.1: Terms using the example of a permanently excited direct current machine.

Sense of rotation

The sense of rotation is the direction of rotation of a machine that the observer sees, when they look at the shaft (congruent to the shaft axis); a direction of rotation that is clockwise is called right-handed rotation, a direction of rotation that is anticlockwise is called left-handed rotation. Three-phase motors have a right-handed rotation, if the exterior conductors L_1, L_2 and L_3 are directed to the terminals U_1, V_1 and W_1 of the terminal board. A reversal of direction of rotation is produced when two exterior conductors are exchanged.

https://doi.org/10.1515/9783110521115-009

There are many different motors:
- Direct current motor
- Synchronous motor
- Asynchronous motor
- Reluctance motor
- Three-phase linear motor
- Starting motor
- Shaded-pole motor
- Universal motor
- Servomotor
- Stepper motor
- etc.

To limit this extensive field, only the most commonly used motors are briefly introduced.

9.1 Transformer

A transformer (see Figure 9.2) serves to convert electrical energy by changing[102] the voltage and the current. In the simplest case (single phase transformer), it consists of one primary[103] and one secondary coil on a joint iron core to increase the magnetic flux density. The function of transformers is based on the induction principle, i.e. the energy is transferred through the magnetic field from the input winding 104 to the output winding by means of induction. Generally the iron core is "lamellated", i.e. made of thin sheet iron[104] to prevent the creation of eddy currents (they would heat up the iron core → loss) in the iron.

The transformation of currents and voltages happens through a different number of turns on the primary or secondary side, which results in the corresponding transmission.

102 In electronics, they are used for converting impedances. Transformers can also be used for the galvanic isolation of circuits (e.g. shaver transformer).

103 Input winding = "primary winding"; output winding (connected to the load) = "secondary winding".

104 The sheet irons are electrically insulated from each other by means of a coating layer. The iron core has to be a magnetic conductor but not an electric conductor. Iron being electrically conductive is not desired when used in transformers. For high-frequency chokes, we therefore use ferrite cores made of iron oxide (haematite – Fe_2O_3) and ceramic additives to get rid of the conductivity of the ferrite in order to prevent eddy currents (and consequently the heat-up of the iron core).

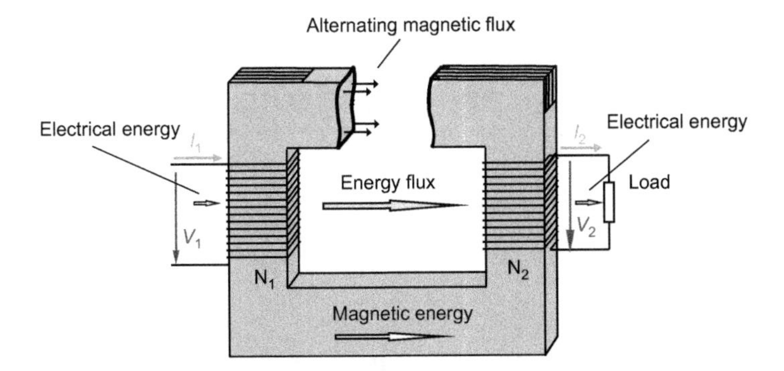

Figure 9.2: Structure of the transformer.

9.1.1 Current transformation

If the output voltage is loaded with a resistance, the current I_2 flows, and at the resistor energy is converted. This energy is obtained from the input side over the magnetic field:

1. In open circuit, magnetising current I_m flows and a magnetic flux Φ_m (main flux) forms in the core.
2. The flux Φ_m is weakened by the counter magnetomotive force $I_2 \cdot N_2$ if it is loaded with current I_2.
3. The now weakened flux Φ_0 results in a smaller self-induced voltage $> V_{10}$ and the *input current* increases. It increases by I_2 so that the additional magnetomotive force generated in the primary coil $I_1 \cdot N_1$ exactly equals $I_2 \cdot N_2$.

The initial main flux Φ_m is restored.

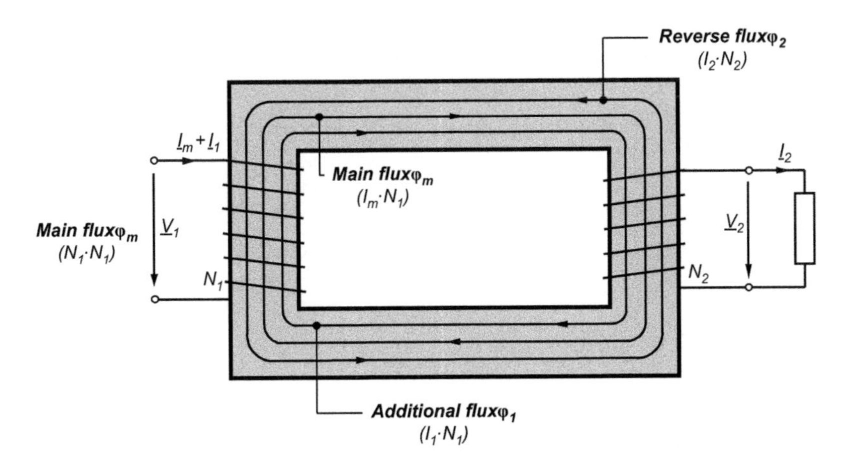

Figure 9.3: Current transformation.

9.1.2 Voltage transformation

The voltage at the primary coil V_1 equals the self-induced voltage V_{10} at every moment (or rather the thereby generated voltage drop) corresponding to the relation:

$$V_1 = -N_1 \cdot \frac{d\Phi_m}{dt} = V_{10} = -L_1 \cdot \frac{dI_1}{dt}$$

The voltage at the magnetically coupled secondary coil V_2 is caused by the changing flux Φ_m (law of induction):

$$V_2 = -N_2 \cdot \frac{d\Phi_m}{dt} = -L_2 \cdot \frac{dI_1}{dt}$$

V_1 Primary voltage in volt
V_2 Secondary voltage in volt
Φ_m Magnetic main flux in Vs or Wb (Weber)
N_1 Number of turns of the primary winding
N_2 Number of turns of the secondary winding
L_1 Inductance of the primary coil in H
L_2 Inductance of the secondary coil in
I_1 Current in the primary coil in A
I_2 Current in the secondary coil in A

The transmission ratio t is formed with the quotient of the voltages:

$$t := \frac{V_2}{V_1} = \frac{-N_2 \cdot \frac{d\Phi_m}{dt}}{-N_1 \cdot \frac{d\Phi_m}{dt}} = \frac{N_2}{N_1}$$

9.1.3 Impedance transformation

Additionally to voltages and currents, impedances are automatically transformed. The transformer transmits the impedances by the square of the transmission ratio:

$$\frac{\underline{Z_1}}{\underline{Z_2}} = \frac{\frac{V_1}{I_1}}{\frac{V_2}{I_2}} = \frac{V_1 \cdot I_2}{I_1 \cdot V_2} = \frac{N_1}{N_2} \cdot \frac{N_1}{N_2} = \frac{N_1^2}{\underline{N_2^2}} \qquad \rightarrow t = \sqrt{\frac{Z_1}{Z_2}}$$

9.1.4 Transmissions – Summary

The primary coil and the secondary coil are located on the same iron core. This causes the main flux Φ_m produced by the primary winding (generated by current I_m)

to also permeate the secondary coil and induce a voltage V_2 through induction. If there is an electrical load on the secondary side, a current I_2 flows, which causes a counterflux Φ_2, which in turn weakens the main flux. The weakening of the main flux is balanced with a current flow I_1 on the primary side. A higher current I_2 causes a higher current I_1 and vice versa.

The following applies for the **transmission ratio** t:

$$t := \frac{V_2}{V_1} = \frac{I_1}{I_2} = \frac{N_2}{N_1}$$

9.1.5 Ideal transformer

A transformer is ideal when its material shows the following properties:

- electrical resistance of the winding material $R_{Cu} = 0$
- electrical resistance of the iron core $R_{Fe} = \infty$
- permeability of the iron core $\mu_{Fe} = \infty$
- permeability of the surrounding air $\mu_{air} = 0$

Because of $R_{Cu} = 0$, the currents of the two windings are absolute inductive reactive currents that contribute to the magnetomotive force of the magnetic circuit. Therefore, **no current heat losses** occur in the windings of an ideal transformer.

Because of $R_{Fe} = \infty$ no eddy currents can flow in the material of the magnetic circuit.

→ **no eddy current losses**

If $\mu_{Fe} = \infty$, the magnetomotive force and thus also the magnetic field strength equal zero in the magnetic circuit which is free of any air gaps. As the magnetic field strength equals zero, it is independent of the level of the magnetic induction prevalent in individual points and no voltage drops can **occur in the magnetic circuit.** Furthermore, there is no hysteresis (see chapter 5.2.3 –Hysteresis loop), hence no **hysteresis losses** can occur.

If $\mu_{air} = 0$, no magnetic stray flux can develop outside the magnetic circuit. Therefore, both windings are penetrated by the same magnetic flux. Furthermore, **no stray-field losses or additional losses** occur.

In the ideal (and therefore lossless) transformer the released complex power S_2 equals the absorbed complex power S_1. **Its efficiency thus is always** 100 %.

9.1.6 Real transformer

With the real transformer the copper windings have a respective resistance larger than zero, as a result, a load-dependent power loss ($P_{Cu} = I^2 \cdot R$) – the so-called **copper losses** P_{Cu} – occur when current flows.

The iron core must always be remagnetised due to the alternating magnetic flux. As the iron core is not an ideal soft magnet (see chapter 5.2.3) so-called core losses occur.

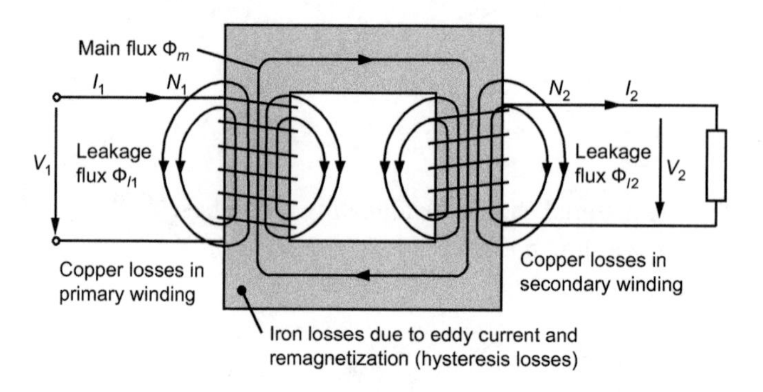

Figure 9.4: Losses in the real transformer.

The alternating magnetic flux induces a voltage in the iron core that drives the eddy currents. These eddy currents heat the iron core and create eddy current losses.

The sum of the eddy current losses and the core losses is called **"iron losses"** P_{Fe}. They are voltage-dependent and frequency-dependent.

Then there also are so-called **additional losses** P_A that occur as dielectric losses (e.g. in the winding insulation) or as eddy current losses in in the components that do not belong to the magnetic circuit (e.g. transformer housing).

Figure 9.5 illustrates the power flow including the losses.

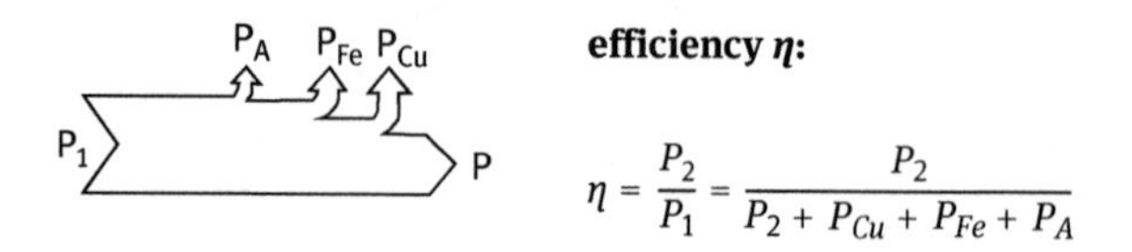

efficiency η:

$$\eta = \frac{P_2}{P_1} = \frac{P_2}{P_2 + P_{Cu} + P_{Fe} + P_A}$$

Figure 9.5: Power flow and losses of energy in the transformer.

P_1 supplied active power
P_2 emitted active power

The consequence of the losses and shortcomings that exist in reality is
- that the transmission applying to the ideal transformer is approximate only, and that experience values (found in tables) need to be used for the construction of transformers.
- that the cross-section of the copper winding must become larger with increasing power of the transformer in order to keep the copper losses small.
- that the lost heat must be dissipated (also note that the volume of a transformer increases by its third power, whereas the surface is only squared; therefore, transformers increasing in size proportionally have less surface for cooling).
- that measures concerning the material and construction have to be implemented to keep the iron losses small (iron must be laminated and as soft magnetic as possible).

9.1.7 Important characteristic values of a transformer

Short-circuit voltage V_{sc}

The short-circuit voltage V_{sc} is determined in short-circuit testing. For this purpose, the secondary side of the transformer is short-circuited, and the input voltage is increased from zero until the rated current I_{1n} flows on the primary side.

The short-circuit voltage is given in % of the rated voltage and referred to as relative short-circuit voltage v_{sc}:

$$v_{sc} = \frac{V_s}{V_n} \cdot 100\ \%$$

V_{sc} Short-circuit voltage in V
v_{sc} Relative short-circuit voltage in %
V_n Rated voltage of transformer in V

Sustained short-circuit current I_{ss}

If you short-circuit the secondary side of a transformer, a sustained short-circuit current forms after the fading of the surge of short-circuit current.

Sustained short-circuit current I_{ss}:

$$I_{ss} = \frac{I_n}{v_{sc}} \cdot 100\ \%$$

I_n Rated current in A
I_{ss} Sustained short-circuit current in A
v_{sc} Relative short-circuit voltage in %

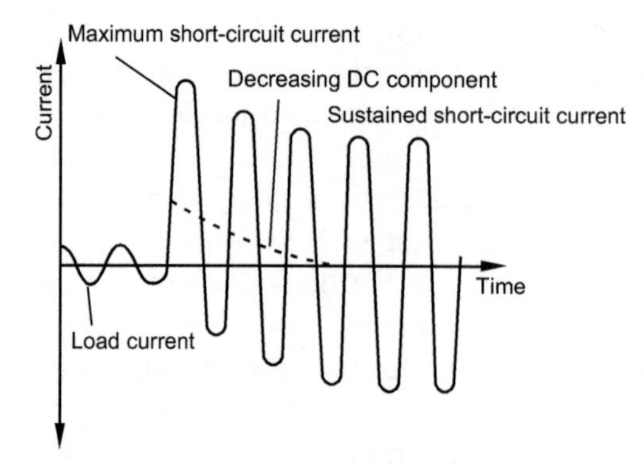

Figure 9.6: Current curve with a short circuit on the transformer.

9.1.8 Types of small transformers

Small transformers have a power rating of up to 16 kVA, input voltages of up to 1,000 V and frequencies of up to 500 Hz.

If a lasting short circuit can remain on the secondary side without the transformer overheating and destroying itself generating a lot of smoke, then the transformer **is deemed short-circuit-proof.**

Normally – with the exception of **autotransformers** – the primary and secondary side of a transformer are always galvanically isolated. That means there is no

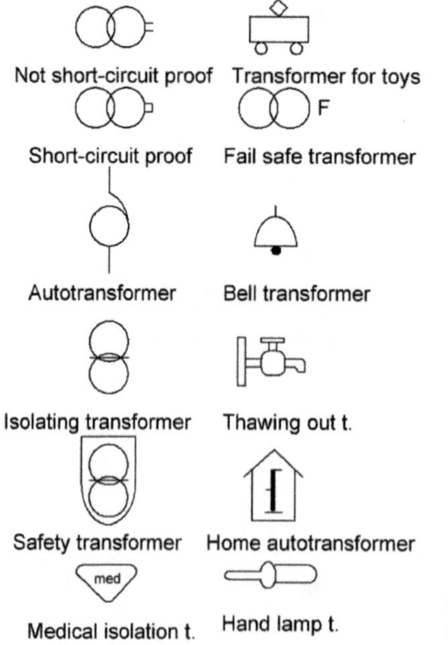

Figure 9.7: Symbols for labelling small transformers.

conductive connection between the primary and the secondary side.[105] A transformer with a transmission ratio of $t = 1$ is referred to as **isolation transformer.**[106]

Transformers are used to create a so-called safety extra-low voltage ($< 25\ V_{AC}$). With safety extra-low voltages, we do not require protection against direct contact (insulation). Such voltages are generated e.g. in **safety transformers in toys**.

9.1.9 Three-phase transformers

Three-phase voltage (as it is used in electric power grids by utilities for energy distribution) can be transformed using three one-phase transformers if their input and output windings are connected in delta or wye connections. In three-phase transformers all windings are applied to an enclosure of steel. Within a three-phase power transformer (as used in substations), the winding has to be cooled. Therefore, the active parts of the transformer (iron core and windings) were put into a metal tank which is filled with mineral oil. The advantages of mineral oil are that, on the one hand, it is electrically insulating, and, on the other hand, it has a high heat capacity. This means it can absorb a lot of heat and release it to the environment if spread on a larger surface.

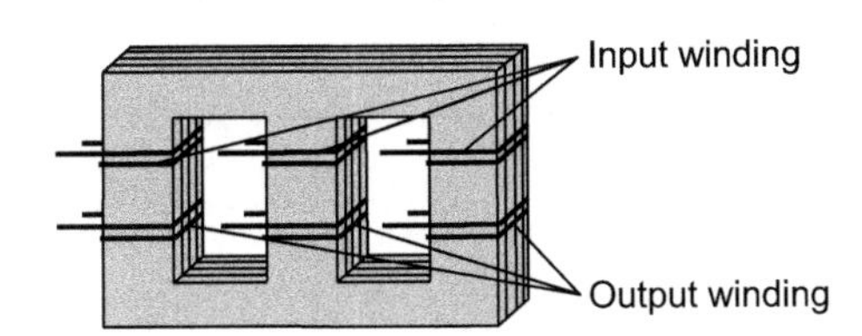

Figure 9.8: Three-legged core for three-phase current.

9.2 Direct current motor (commutator motor)

9.2.1 Structure and functioning

The direct-current-carrying excitation winding ($F_1 - F_2$) generates the excitation magnetic field, which closes over the armature (see Figure 9.9: Externally excited DCM (anti-clockwise rotation) Figure 9.9). If the armature contains a current-carrying

105 The energy is transferred from the primary to the secondary side only by means of the alternating magnetic flux.

106 Isolation transformers are used in measurement technology if the voltage is to be decoupled from the earth potential. There is no voltage to earth since the secondary side is not earthed. Therefore, these transformers constitute a safety advantage that is used e.g. in shaver supply units.

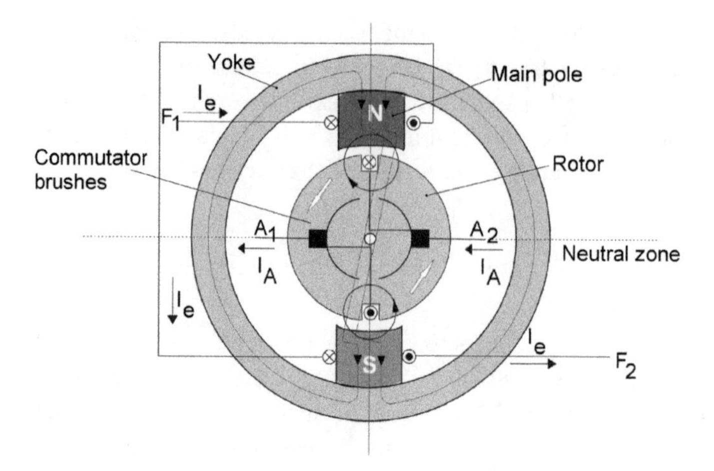

Figure 9.9: Externally excited DCM (anti-clockwise rotation).

conductor loop, the magnetic field of the conductor loop interferes with the excitation magnetic field below every main pole. At every pole, a force affects the conductor loop. The direction of the force is determined by the rule for motors.

A torque is generated that moves the conductor loop into the neutral zone where the torque is then reduced to zero. Due to the kinetic energy, the armature crosses the neutral zone. As soon as it has passed this zone, its polarisation is reversed by a commutator, which leads to an ongoing rotation. To create a constant and high torque, we substitute the conductor loop with several coils within the armature.

Direction of rotation:	For the reversal of direction of rotation the polarity of the armature current is reversed.
Starting DCMs:	By means of starting resistances the starting current can and has to be limited.
Speed control:	Through increasing the armature voltage (e.g. using an electrical rectifier) the rotational speed of the motor is increased up to the rated rotational speed. Alternatively, the excitation field can be weakened →"field weakening": In case of the rotational speed increasing too greatly, the electrical rectifier and armature can be destroyed due to the centrifugal forces: "The motor runs away."

9.2.2 Types of direct current motors

Direct current motors can be distinguished through the connection of excitation winding and armature. The two main types produced are externally excited motors and commutated series-wound motors.

Table 9.1: Connections of DC motors.

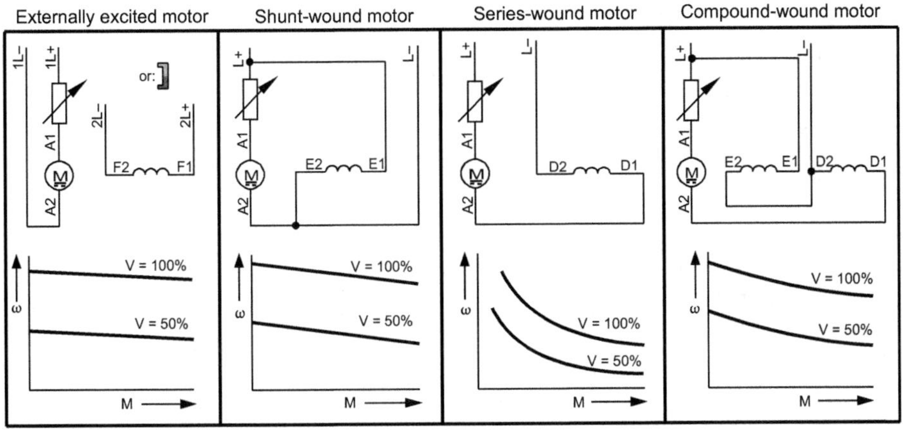

Externally excited motor	Shunt-wound motor	Series-wound motor	Compound-wound motor

Externally excited motor

In externally excited motors, the armature circuit and the excitation winding circuit are supplied by two independent voltage sources. For motor powers up to about 10 kW, the excitation happens through permanent magnets. To start the motor or reduce the rotational speed, the armature voltage is reduced e.g. through a starting resistance. A rotational speed higher than the rated rotational speed is reached due to field weakening (e.g. using a field regulator). With externally excited motors, the rotational speed only slightly decreases under load. They do not run away at idle.

Areas of application: Engines with a high rotational-speed control range e.g. machine tools, milling machines, conveyor systems.

Series-wound motors

In series-wound motors, armature and excitation winding are connected in series. The armature current is, at the same time, the excitation current. During start-up and under load, high current consumption leads to a high armature field and a high excitation field. Commutated series-wound motors, hence, have the highest torques. In open circuit, the excitation field is continuously weakened. This can lead to an increase in rotational speed up to the destruction of the motor. The rotational speed of commutated series-wound motors is very much dependent on the load; they **run away at idle.**

Areas of application: Commutated series-wound motors are especially used in electric vehicles such as trams, electric carts and forklifts. If rotor and stator are laminated, operating the motor with alternating current is possible (universal motors).

Advantages and disadvantages of direct current motors (DCM):
+ DCMs must have a great starting torque and can (or rather must[107]) start-up under load
+ high starting torque
+ simple control of rotational speed (n) and load torque (M)
+ good concentricity properties
+ high rotational speed possible
- brushes necessary

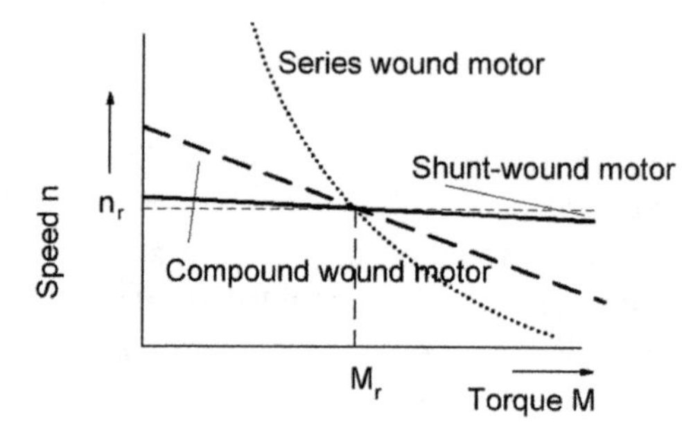

Figure 9.10: Behaviour of rotational speed of direct current motors.

9.3 Three-phase asynchronous motor

Asynchronous motors are the most important three-phase motors. The stator rotating field induces a voltage in the rotor that drives a current. Therefore, the rotor rotates. This is the reason why asynchronous motors are also referred to as induction motors. Asynchronous motors require a slip for the induction of the rotor current.

Structure

The stator consists of the housing, the stator stack of electrical steel laminations and the stator winding. The beginnings and endings of the coils are connected to the terminal board. The rotor consists of the stack of electrical steel laminations on the shaft and of the aluminium or copper conductor bars inserted into the grooves. At the front side of the stack, the conductor bars are connected with shorting rings. Conductor bars and shorting rings form the rotor winding and look like a cage (squirrel-cage rotor).

107 Commutated series-wound motors run away at idle!.

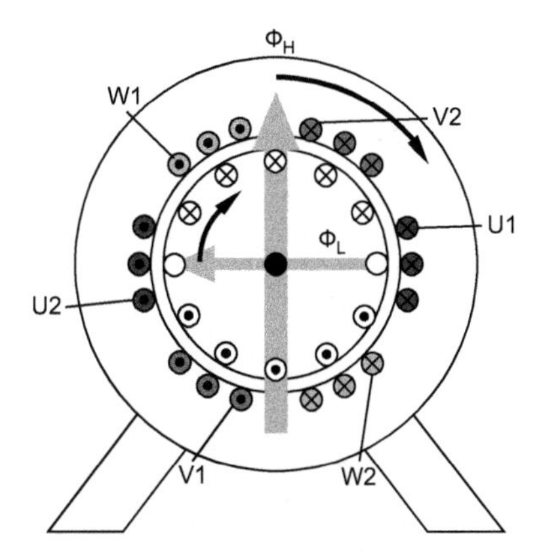

Figure 9.11: Three-phase asynchronous motor with squirrel-cage rotor.

Mode of action: The cage winding can be seen as the simplest form of a three-phase winding. The squirrel-cage induction motor at starting torque corresponds to a transformer. At first, the rotating field in the stator winding causes a flux change in the conductor loops of the stationary rotor. The rate of flux change is proportional to the rotating-field speed. The induced voltage drives the current through the conductors that are connected to the shorting rings. The current-carrying conductor in the magnetic field is affected by the Lorentz force which causes a rotation of the rotor.

9.3.1 Generation of a rotating field

If we turn a bar-shaped permanent magnet or an electromagnet round its axis, a rotating magnetic field is generated. This is also the case with three coils displaced by 120° (see Figure 9.12 a) that are passed through by three-phase current. In motors, the coils are evenly distributed in form of windings in the grooves of the stator (see Figure 9.12 b).

Each of the coils creates a magnetic alternating field. The magnetic fields of a coil interfere with each other, which leads to the generation of a resulting field. The position of said field depends on the instantaneous values of the currents phase-shifted by 120°. With three coils, a two-pole rotating field is generated; with six coils that are displaced by 60°, respectively, a four-pole rotating field is created.

The rotating-field speed n_S is determined using the mains frequency f and the number of pole pairs p:

$$n_S = \frac{f}{p}$$

n_S Rotating-field speed in Hz
f Mains frequency in Hz
p Number of pole pairs $[p] = 1$

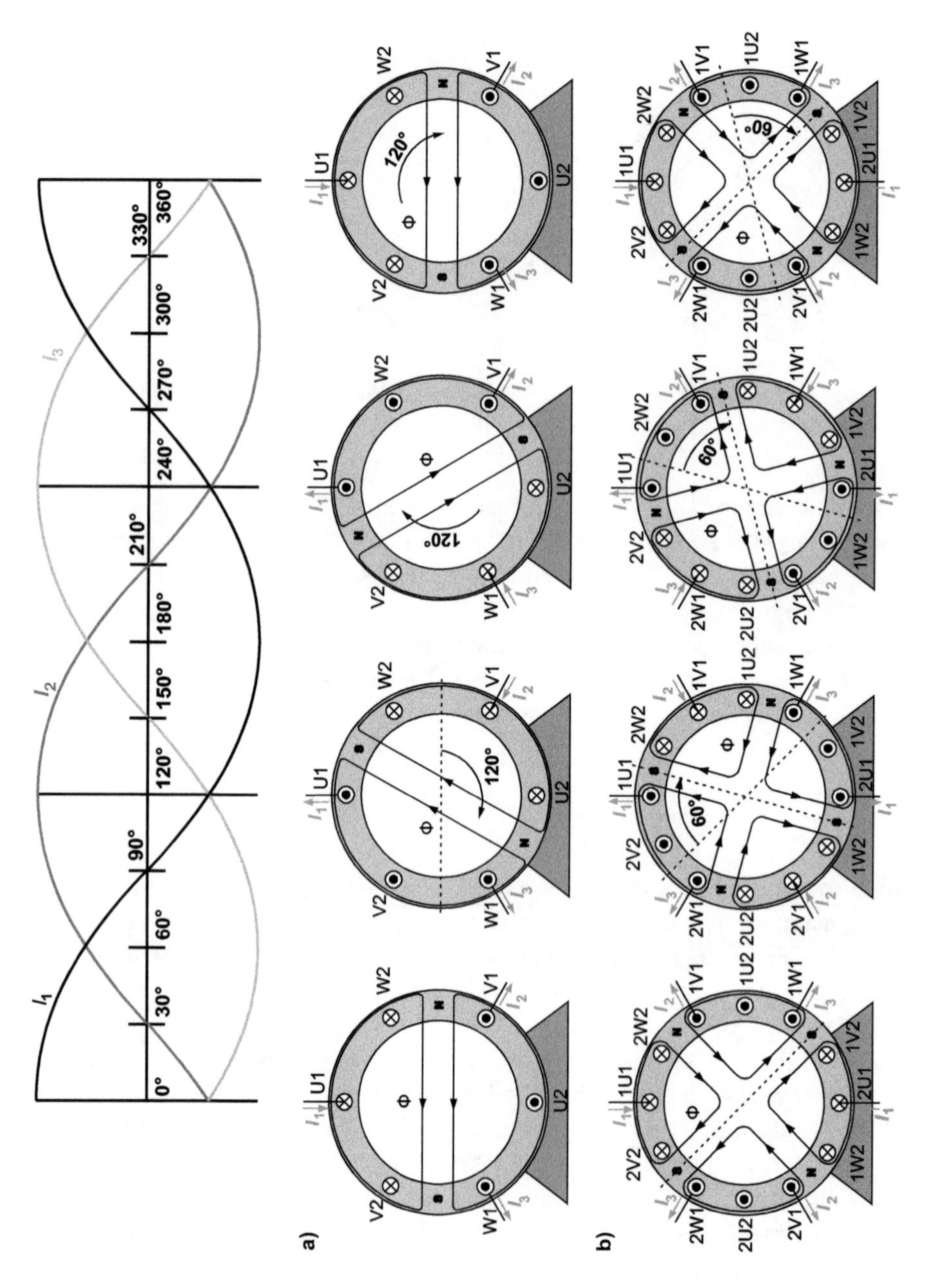

Figure 9.12: Generation of a rotating field a) two-pole rotating field; b) four-pole rotating field.

9.3.2 Power *P* and torque *M*

Two of the most important characteristic quantities in motors are their power and their torque. Motors convert electrical energy provided by the supply network into mechanical work; generators convert mechanical work (drive) into electrical energy. By measuring the force at the circumference of the driving disc of the motor we can determine the generated torque. Eddy current brakes, magnetic powder brakes or pendulum machines measure these torques.

The torque M is given by

$$M = F \cdot r$$

The emitted power P_2 is given by

$$P_2 = M \cdot \omega$$

considering that the angular velocity ω is calculated using $\omega = 2 \cdot \pi \cdot n$.

M Torque in Nm
F Force in N
r Radius in m
P_2 Power output in W
ω Angular velocity in $rad \cdot s^{-1}$
n Rotational speed in s^{-1}

For the power consumption P_1 in three-phase motors, the following applies:

$$P_1 = \sqrt{3} \cdot V \cdot I \cdot \cos\varphi$$

P_1 Power consumption in W
V Conductor voltage in V
I Conductor current in A
$\cos\varphi$ Active power factor

9.3.3 Efficiency η

In machines, active power losses happen in form of heat. The part that forms due to magnetic reversal and eddy currents is called **iron loss.** The part that is caused by the current flowing through the active winding resistances is called **winding loss.** Further losses are caused by the fan and the friction in the bearings and the brushes (if there are any). A measure for the occurring overall losses is the efficiency η. This is the quotient between released and absorbed power:

$$\eta = \frac{P_2}{P_1}$$

η Efficiency
P_1 Power consumption in Watt
P_2 Power output in Watt

9.3.4 Slip *s*

According to Lenz's law the magnetic field caused by the rotor current forms a torque that rotates the rotor into the direction of rotation of the stator rotating field. If a rotor reaches the rotating-field speed, the flux change in the conductor loops would be zero, thus the induced voltage and the torque causing the rotation would also equal zero. The rotor speed is hence always slower than the rotating-field speed (asynchronous). The difference of rotational speeds is called slip speed Δn.

$$\Delta n = n_S - n$$

$$s = \frac{n_S - n}{n_S} \cdot 100\ \%$$

Δn Slip speed in s^{-1}
n_S Rotating-field speed, synchronous rotational speed in s^{-1}
n Rotor speed in s^{-1}
s Slip in %

The slip of asynchronous motors depends on the load and, for the power rating, is about 3% to 8% of the rotating-field speed.

9.3.5 Torque curve

The torque curve indicates the increase of the torque until the greatest motor torque, the **tilting moment M_T**. When starting the motor, the moment is called **starting torque M_S**.

With greater rotor speed the rotor induction current is smaller due to the smaller slip. This causes a decrease of the motor torque.

The **pull-up torque M_P** is the smallest motor torque after the start-up. With the rated rotational speed n_n the motor has the **rated torque M_n**. In open circuit it almost reaches the **rotating-field speed n_S**.

The rated rotational speed of the motor lies in the range where the torque curve starts dropping towards the growing rotational speed. During a load surge the rotational speed decreases but the torque of the motor increases, thus the rotational speed increases as well yet again. → rotational speed remains stable.

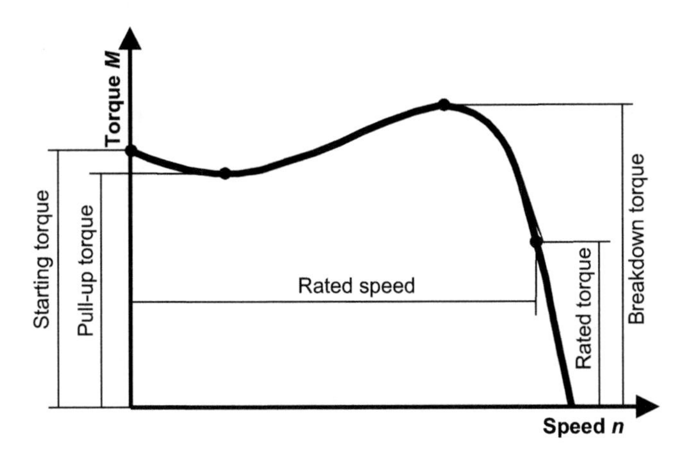

Figure 9.13: Torque curve of a squirrel-cage induction motor.

9.3.6 Starting squirrel-cage induction motors

To decrease the current during start-up ("breakaway starting current") and thus to prevent an overload of the electrical installation, a start-up procedure has to be implemented for asynchronous motors with greater power.

There are:

- Stator starter with resistances: Active resistors or choke coils in the motor supply decrease the stator voltage.
- Soft-starter circuit for squirrel-cage induction motors are used in smaller motors. The limiting of the breakaway starting current is achieved only through a resistor in an exterior conductor.
- Starting transformers: The stator voltage is reduced using an autotransformer.
- Wye-delta starting method: The windings are connected in wye connection (high torque at low speed) and only changed to delta connection (lower torque at higher speed) after the start-up.
- Electronic soft-starters: With thyristors connected in anti-parallel used as supply for the motor, we can reduce the stator voltage for start-up through phase-fired control (PFC).

Wye-delta starting method

The wye-delta starting method is the one most commonly used in households and in the industries to limit the breakaway starting current. By means of a (usually already attached) manual wye-delta switch upstream of the motor, the breakaway starting current is limited. Thereby the windings of the motor are first operated with a wye connection and, after reaching the rotational speed, operation is switched to a delta connection. These starters are also available as wye-delta contactor circuit, where the switchover to delta occurs automatically (e.g. after a certain amount of time or when reaching a certain rotational speed) (see Figure 9.14).

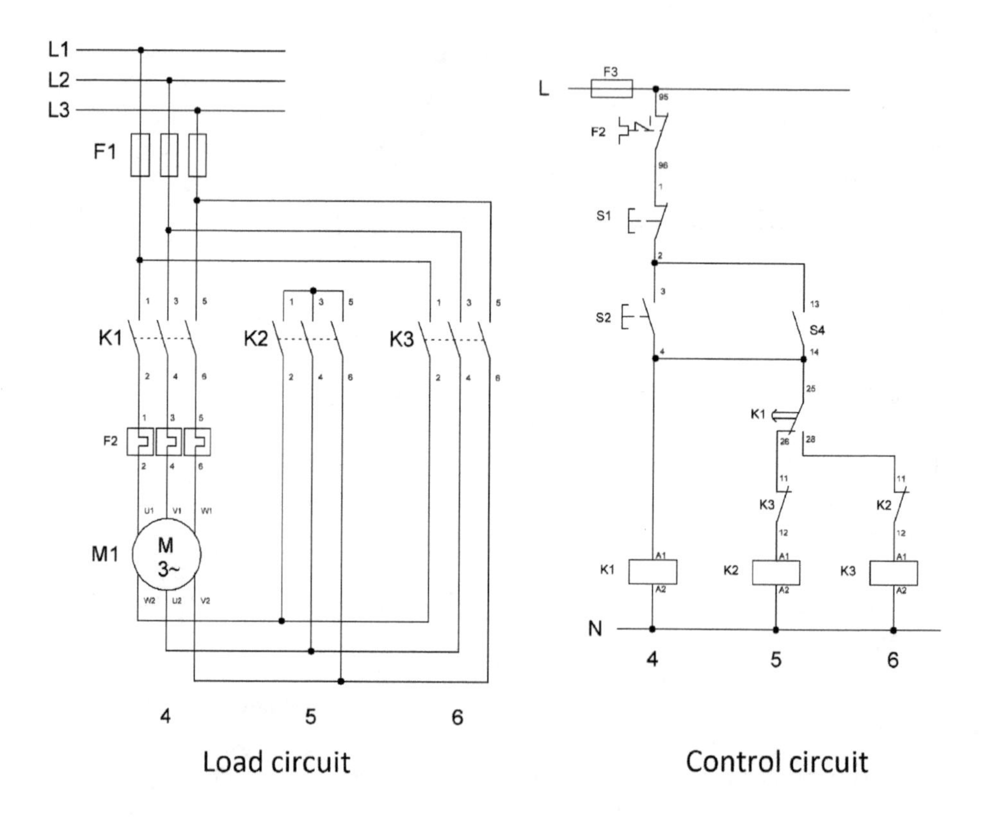

List of operating equipment:

F1 Main fuse
F2 Thermorelay as motor protection
K1 Mains contactor
K2 Wye contactor
K3 Delta contactor
M1 Three-phase motor
F3 Safety fuse for the control circuit
S1 Push-button Off
S2 Push-button On

Figure 9.14: Automatic wye/delta switchover (solved representation).

Circuit description (see Figure 9.14) When pressing the push-button S2 the mains contactor K1 is activated via current path 4. Simultaneously, the wye contactor K2 is activated via the delayed contact K1 via current path 5. The contactors K1 and K2 maintain the current path 5 via the contact K1. The motor starts in wye connection. After a preselected time, the delayed contact of the contactor K1 switches from 25/26 to 25/28. K2 drops and contactor K3 is activated (disabling contact K2 11/12). If the push-button S1 or the excess current relay F2 opens, all contactors deactivate.

9.3.7 Applications of squirrel cage induction motors

Squirrel cage induction motors are cheap to produce, easy to maintain and free of radio interference. Standard motors have powers of approximately 200 kW. They serve as drive in e.g. saws, lifting gear, blowers and machine tools. Operating with the maximum rated load is the most economic way (see Figure 9.15). The motor is highly efficiency and has a great active power factor if it works at the maximum rated load.

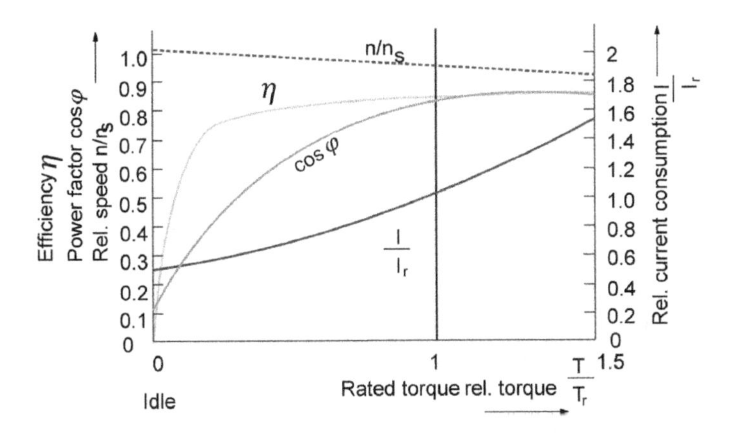

Figure 9.15: Typical characteristic curve of squirrel-cage induction motors with powers of approx. 2 kW to 5 kW.

9.4 Synchronous motor

Structure

The stator of the synchronous motor is structured in the same way as in the asynchronous motor. In the stator stack of electrical steel laminations, there is a three-phase winding used to generate the rotating magnetic field. The rotor with a solid or laminated pole core has an excitation winding that is supplied with direct current through the slip rings. It acts as pole wheel (electromagnet) and its number of poles equals the number of poles in the stator winding. In smaller motors, also permanent magnet rotors are used.

Mode of action

As soon as the motor is turned on, its stator rotating field is immediately in accordance with the number of poles and the mains frequency. The rotor, however, has the rotational speed zero. Due to inertia, it is attracted by the opposite poles of the stator rotating field but is repelled by the like poles of the field shortly afterwards. The pole wheel cannot immediately follow the rotating-field speed. Therefore, most synchronous motors have an additional short-circuited winding as starting traction control so that the synchronous motor can start-up as asynchronous motor and can synchronously rotate with the starting traction control at approximately the same rotating-field speed. During start-up, the excitation winding must be closed by means of a resistor so that the induced voltage does not destroy the insulation of the winding. The short-circuited winding also has the advantage that it softens the oscillation of the rotor in case of load impacts during operation. Consequently, this winding is also referred to as damper winding.

Operating behaviour

During operation the rotor of the synchronous motor always rotates with the rotational speed of the rotating field – therefore synchronous to it (hence the name "synchronous motor"). Under load the distance between the poles of the pole wheel and the poles of the rotating field increases – the pole wheel lags behind the rotating field by the load angle ϑ, compared to its position in open circuit (without load). If the load and therefore the load angle are so big that the pole wheel is located in between two poles of the stator, the torque is at its maximum (tilting moment); the pole of the rotating field running ahead in the direction of rotation pulls the pole wheel and the lagging pole pushes it. Under load at the tilting moment the rotor falls "out of step" (rotational speed of the pole wheel does not correspond to the rotational speed of the rotating field anymore) and would stop without damper winding.

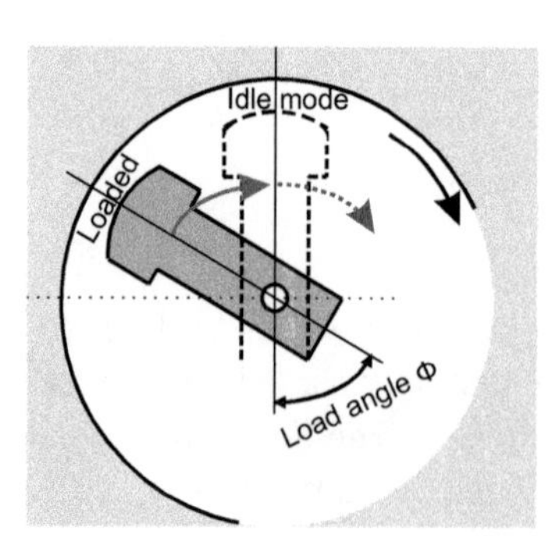

Figure 9.16: Load angle in a synchronous motor.

Application

Because of their constant rotational speed, synchronous motors are also used as small alternating-current motors with a permanent magnet rotor.

9.5 Synchronous generator

Electrical energy for the power grid is mainly generated by synchronous generators. In power plants, especially inner pole machines are used, as the excitation energy that is supplied to the pole wheel through slip rings is small compared to the energy generated in the stator windings.

Structure

The rotor carries the excitation winding which is supplied with direct current through slip rings. As the rotor field consequently does not change, the rotor is usually manufactured from massive steel (not laminated). Rotors designed for lower rotational speed have pronounced poles (pole wheel). Rotors for high rotational speed usually are two-pole rotors and are constructed as non-salient pole rotors (="turbo rotors").

The excitation current can be obtained from the grid with a rectifier. The excitation current supply can also be provided by an excitation set coupled to the generator shaft for self-excitation. The permanent auxiliary generator excites a three-phase main excitation machine. The voltage is rectified and supplied to the excitation winding of the generator. The stator is arranged in layers of electric sheet metal. A three-phase winding is located in the stator grooves. Stator and rotor have the same number of poles.

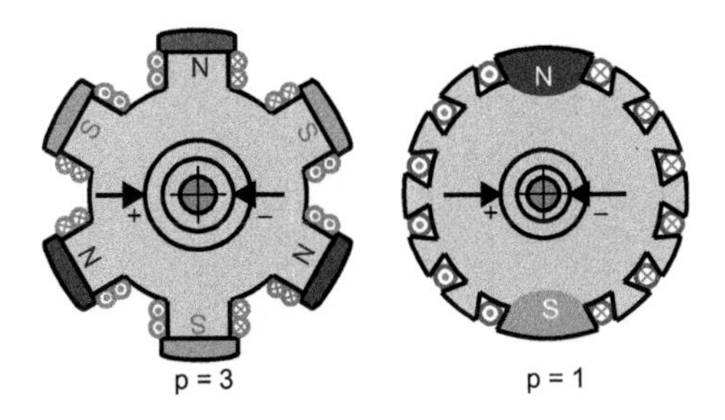

Figure 9.17: Types of rotors of synchronous generators: a) pole wheel, b) non-salient pole rotor.

Mode of action and operating behaviour

The rotor is powered by a power engine (e.g. turbine). The direct current in the excitation winding creates a stationary magnetic field related to the rotor. Due to the rotation of the rotor, a rotating field is generated for the stator that induces three voltages in the three strings of the stator winding which are phase-shifted by 120°. The stator delivers three-phase current.

The level of induced generator voltage depends on the excitation current and on the rotational speed of the rotor. Since the frequency is often preset, it determines the rotational speed of the pole wheel. The voltage is adjusted through the excitation current. With saturation of the pole core, the open-circuit characteristic curve flattens. Also, without an excitation current, voltages are generated through pole wheel remanence. Said currents can be as high as several 100 V in high-voltage generators.

The voltage generated by synchronous generators increases with the rotor speed and the excitation current. The frequency is a result of the rotor speed.

9.6 Stepper motor

Structure

Figure 9.18 illustrates the general structure of a stepper motor. The stator, in this representation, has two direct-current excitation windings E1 and E2, while the rotor is a permanent magnet rotor. The excitation windings are – in order to demonstrate the operating principle – supplied with direct voltage via the two switches S1 and S2. The excitation windings develop a north and a south pole according to the direction of the current flow. The magnetic flux is supplied to the pole pieces through the iron[108] cores.

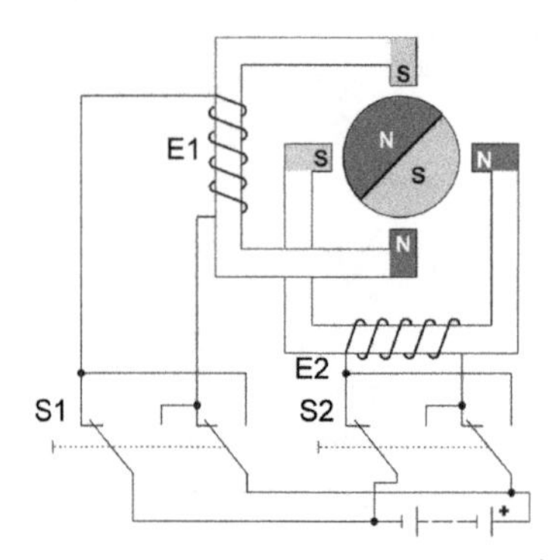

Figure 9.18: General structure of a stepper motor.

Mode of action

At the switch position illustrated in Figure 9.18, respective common north and south poles of the two excitation windings form, which are located between the pole pieces of the two excitation windings. The permanent magnet rotor adjusts its

108 In contrast to air, iron is an extremely good magnetic conductor.

poles to the windings (see Figure 9.18). If switch S2 is actuated, the polarity of the excitation winding E2 changes. Thus, following the new poles of the excitation windings, the rotor turns clockwise by 90° (see Figure 9.19).

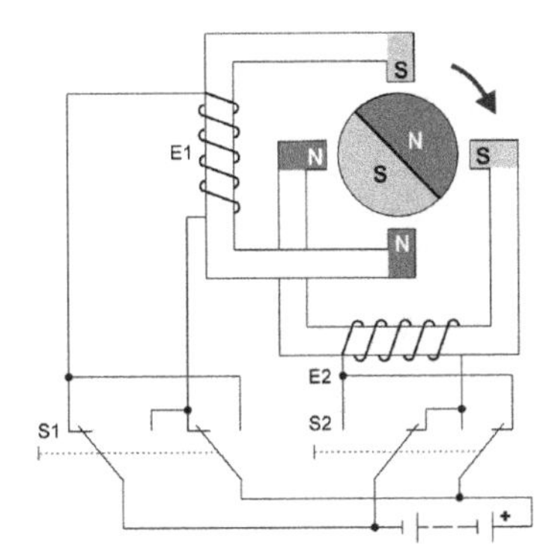

Figure 9.19: Stepper motor after actuating S2, p = 1, m = 2 → $\alpha = 90°$.

If switch S1 is now actuated, the polarity of the excitation winding E1 is reversed and the rotor engages in a new position after a rotation of further 90°. In case of further switchovers with S2 and S1, the rotor performs the according rotations. The respective rotation is referred to as step angle. This angle decreases, the more phases m occurs and the higher the number of pole pairs in the motor is.

$$\text{Step angle } \alpha: \qquad \alpha = \frac{360°}{2 \cdot p \cdot m}$$

As mechanical switches are subject to wear and tear and allow only low switching speeds, stepper motors are in reality operated with electronic control switches (designed as complete control electronics).

Stepper motors thus need a special kind of control electronics, where the rotor can be operated with a stepwise but also with a steady rotation. The small step angle requires a high number of poles in the motor; this requires a specific arrangement due to step angles < 7.5° ("homopolar motor principle").

Application: Up to a power of 1 kW, stepper motors can be used economically for positioning tasks. (e.g. dot-matrix printer, adjustment of valves in heating system)

9.7 Review questions

?

1) Transformers:
   a) Structure and operating principle of a (one-phase) transformer?
   b) Why do transformers "hum" and what is the frequency of the "humming"?

c) In what way does the drawn current change in an ideal transformer, if the load current is doubled?
d) What are the current conversion ratio, voltage conversion ratio and impedance conversion ratio of an ideal transformer?
e) What is the complex power S_1 of an ideal transformer, if it delivers the complex power S_2?
f) What losses occur in a real transformer und how do you minimise them or ordeal with the consequences?

2) Describe the principle and the types of direct current motors.
   a) What is the fundamental difference between an externally excited motor and a commutated series-wound motor?
   b) What needs to be considered during the speed control process field weakening?
3) Three-phase asynchronous motors: Describe the structure, the operating principle, the slip speed and the torque curve.
   a) In what range of the torque curve is the rated rotational speed of an asynchronous motor located? Why is it located in this range?
   b) Why is there generally a slip in an asynchronous motor?
4) Synchronous motors
   a) Structure and operating principle? Difference to the asynchronous motor?
   b) Examples of use?
5) Synchronous generators?
   a) Structure and operating principle?
   b) Why are synchronous generators for large-scale power generation designed as internal-pole generator?
6) Stepper motor: Structure and operating principle?

9.8 Exercises

EXERCISE 9.1

A three-phase transformer with a rated power of 250 kVA has a maximum efficiency of 98.5 % at a load of 150 kW. Calculate the iron losses, the copper losses at rated power, the efficiency at rated power and $\cos\varphi = 1$. How big is the efficiency over a duration of one year when the transformer is loaded 2000 h with rated power, 6000 h with 60 % of rated power and the remaining time with 20 % at a $\cos\varphi = 0.9$ inductive load?

EXERCISE 9.2

A three-phase asynchronous motor with 4 poles for a frequency of 50 Hz has a nominal rotating speed of 1440 min^{-1}. Calculate the slip.

Exercise 9.3

On the power rating plate of a three-phase asynchronous motor with 110 kW following specification is given: 400 V, 197 A, cos $\varphi = 0.86$, 1485 min^{-1}. Calculate the nominal efficiency.

Exercise 9.4

Calculate how much the initial current can be reduced by applying the Wye-delta starting method?

10 Dangers of electricity

10.1 Direct effects on humans

Electric current causes various reactions in the human body that can be categorised in three main groups:
- Thermal effects
- Chemical effects
- Muscle-contracting or paralysing effects

The effects on humans are determined mainly by the type of current (AC or DC), the duration of the impact and the current path through the body.

The heat exposure due to electric current leads to burns at the points of entry and exit in case of high current intensity. The resulting electric arcs may lead to the charring of body parts. As a consequence of severe burns, the kidneys may be overburdened which can lead to kidney failure and ultimately to death.

Body fluids such as perspiration, saliva, blood and cellular fluid contain mobile ions and are therefore electrolytes, which mean that they are electrically conductive. The chemical effect of electric current, especially in case of a long exposure time, can lead to the electrolytic decomposition of blood which, in turn, can lead to severe poisonings. Such a secondary disease may also occur several days after the incident and is therefore especially treacherous.

Almost all human organs function with the help of electric impulses generated by the brain. Weak voltage pulses of approximately 50 mV control muscle movements in our bodies. The pulses are transmitted from the brain to the muscles through nerves. In case of an externally generated current flow through the body, the muscles might cramp. If the hand is affected, one might not be able to let go of a grasped object. If the chest is affected, respiratory arrest occurs. Furthermore, this may cause cardiac arrest or ventricular fibrillation – the heart quivers instead of pumping due to disorganised heart muscle movements.

A dangerous current flow is the result of a voltage that drives the current through the body. In order to determine what level of contact voltage can cause a fatal accident, one needs to consider the following aspects:

10.1.1 Resistance of the human body, dangerous contact voltage

The fatal voltage level for persons can be determined by the maximum electric current and the resistance of the human body-. the. The extremities have a resistance of approx. 500 Ω, the torso is considered resistanceless (see Figure 10.1).

https://doi.org/10.1515/9783110521115-010

From experience, it is well-known that a current of 50 mA can be fatal if the current passes through the heart. In that case, the points of entry and exit of the current are on two separate extremities. The resistance of the human body is hence estimated to be 1,000 Ohm. The level of voltage at which electric current becomes life-threatening is calculated in accordance with Ohm's law ($V = I \cdot R$):

$$50 \cdot \underbrace{10^{-3} \cdot 1 \cdot 10^{3}}_{1} = 50\ V$$

It is the current which constitutes danger. The voltage, however, is responsible for the current flow and is often known or easily measurable. Therefore, it makes sense to be aware of a voltage rating that can cause dangerous current.

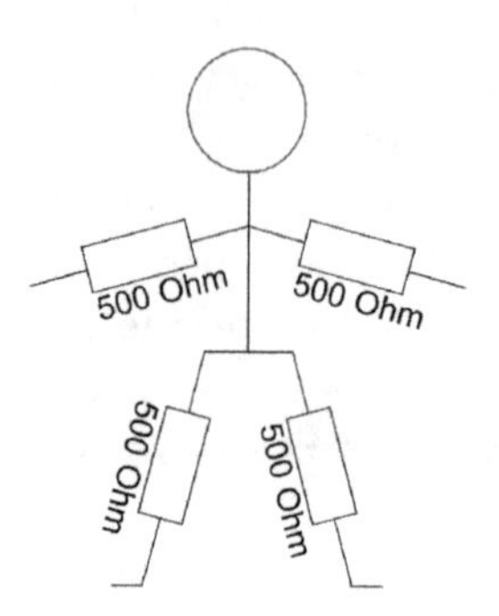

Figure 10.1: Resistance of the human body.

Mortal danger:

- Alternating voltage: **above 50 V** for humans, for animals starting at 25 V
- Direct voltage: **above 120 V** for humans, for animals starting at 60 V

Electrical installations in buildings have to be carried out ensuring that, in case of fault, contact voltage remains below 50 V. This can be guaranteed through insulation, electrical partition, earthing of metal housings and with residual-current devices that interrupt the current flow within a very short amount of time. Table 10.1 lists examples for voltages that are present in our everyday life.

The wire of an electric fence delivers voltage pulses > 4000 V to earth. It does, however, not constitute a mortal danger because the voltage source is equipped with a high internal resistance preventing the flow of dangerous current.

Table 10.1: Examples of electric voltages.

Example	Voltage
Three-phase supply in the home	230 V / 400 V AC, 50 Hz
Car battery, truck battery	12 V DC, 24 V DC
Tram in Graz	600 V DC
Overhead line; Austrian Federal Railways (ÖBB)	15 000 V AC, $16\frac{2}{3}$ Hz
Thermoelectric voltage	40 µV DC
Overhead power line (maximum voltage)	380 000 V AC, 50 Hz
Lightning[109]	up to 1 000 MV

To assess whether a current constitutes danger, it is crucial to not only consider the intensity but also the **exposure time**. This correlation is represented in Figure 10.2.

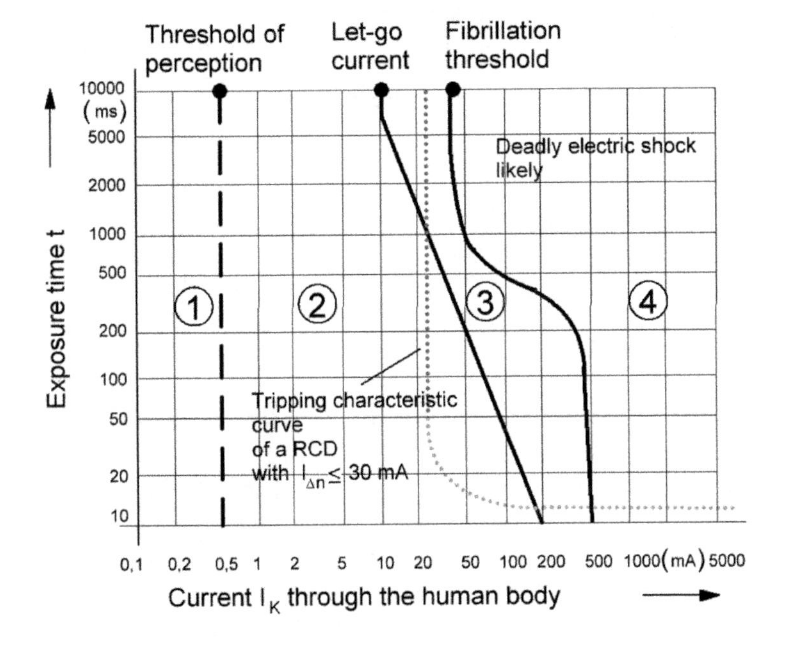

Figure 10.2: Effective ranges of alternating current of 50 Hz on adults.

109 Despite the high lightning current, two thirds of victims survive a direct lightning strike. The reason for this is that the majority of lightning current is discharged through surface discharge via the body surface (skin effect).

ϟ With the effective ranges depicted in Figure 10.2, the following impacts are to be expected:

Table 10.2: Impacts of electric current depending on current and exposure time.

Level	Impact
1	Even with an arbitrarily long exposure time, **effects or reactions remain below the threshold of perception.**
2	**Range of perception: Humans perceive currents starting at 2 mA as more or less unpleasant.** Startle responses can lead to secondary accidents such as falling. Up to **10 mA (=let-go threshold)** an arbitrarily long exposure time is still non-hazardous.
3	Currents above the let-go threshold lead to **muscle contractions and respiratory problems.** If a person grasps a wire under voltage, it may happen that they cannot release their grasp anymore due to cramping of muscles. If the diaphragm or the respiratory centre of the brain is affected, respiratory paralysis – resulting in unconsciousness – occurs due to prolonged exposure time. Reversible disorders of impulse production and impulse conduction of the heart are possible. Danger of ventricular fibrillation.
4	High probability of ventricular fibrillation. Cardiac arrest, respiratory arrest and severe (inner) burns occur with increasing current and length of impact. **A lethal impact of current is likely at this level.**

A residual-current device serves as protection for people and activates in practice after 20 to 30 ms. Hence, it protects the user of an electrical device from harmful impacts of current.

10.1.2 Residual-current device (RCD)

Residual-current protection devices disconnect the circuit from all terminals if a fault current that is higher or as high as the **rated residual current** $I_{\Delta n}$ occurs. RCDs thus contribute to the reduction of life-threatening accidents involving electric current. They consist of a ring-shaped iron core through which all currents pass in the case of faultless operation. In that case, the directionally dependent sum of currents passing through the device equals zero at any given moment; the contacts remain closed. When a fault occurs, the load current or a part of the load current does not pass through the RCD anymore. Consequently, the sum of currents in the residual-current device does not equal zero; voltage is induced in a secondary winding and the device uses an interrupting mechanism to disconnect the circuit from all terminals. The RCD is equipped with a test button that causes a pre-defined fault current when activated. In this way, the correct operation can be tested. The most important parameter is the rated residual current $I_{\Delta n}$, which triggers the RCD within the trigger time (< 300 ms).

The maximum of allowed rated residual current $I_{\Delta n}$ for the protection of people is 30 mA. Residual-current devices are divided into groups according to the type of fault current they are able to detect. Figure 10.3 illustrates the protective effect and the purpose of an RCD. The bold line depicts the current path in case of fault with a connected protective conductor. The bold dashed line depicts the current path passing through a person if the protective conductor is interrupted. The protective conductor, hence, additionally protects against contact voltage in case of fault.

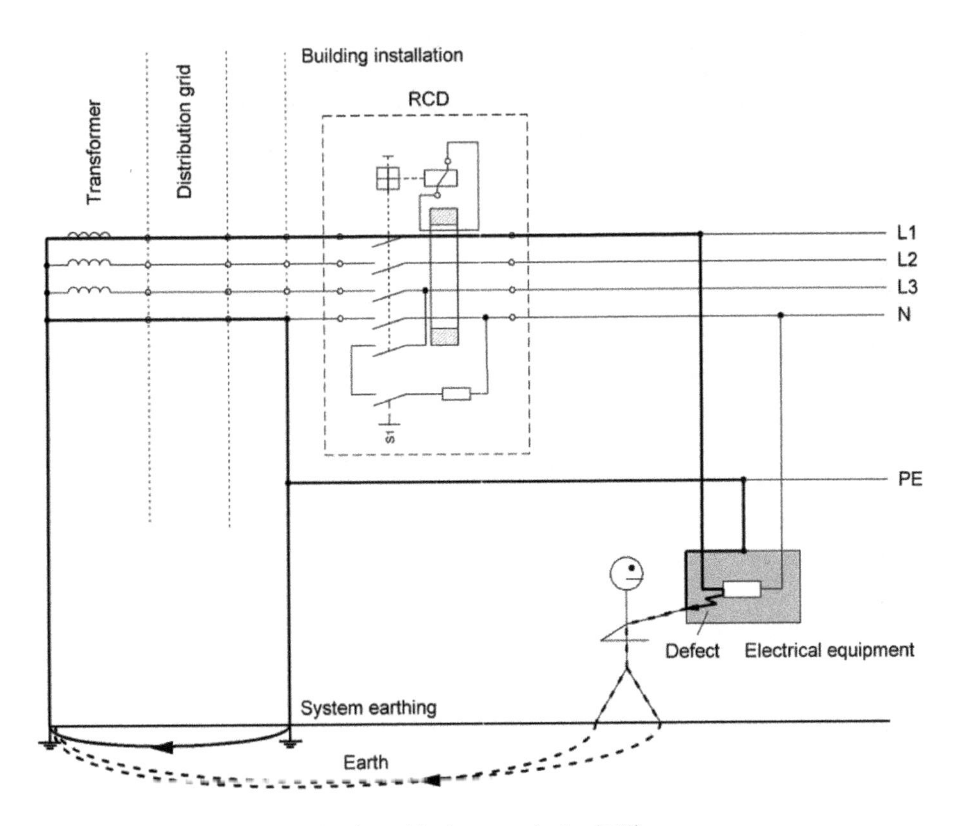

Figure 10.3: Functional schematic of a residual-current device (RCD).

10.2 Indirect impact on humans

Electric current has four main effects:

- Heating effect
- Lighting effect
- Magnetic effect
- Chemical effect

High currents can endanger the health or security of people and/or objects. An example are overheated lines in an electronic device, caused by too high current, can, in consequence, cause a fire. Current can produce poisonous or dangerous gases through electrolysis. A strong magnetic field caused by current flow can destroy or influence electronic equipment. An electromagnetic field can disturb radio communication.

A spark occurs in case of electrostatic discharge (charge caused by e.g. static electricity) or during switching operations.[110] An electrical spark can cause chemical reactions or explosions (e.g. dust explosion) and can create broadband electromagnetic interference signals. In rooms where there is an explosion hazard (e.g. fuel laboratories, chemical storages, dust exposed rooms), open sparking must be absolutely avoided. People with a cardiac pacemaker must avoid strong magnetic fields (e.g. MRI, induction heating) as they might cause interferences with the pacemaker.

10.3 Types of faults, voltages in case of fault

Figure 10.4 shows different types of faults. Figure 10.5 illustrates the difference between the terms fault voltage and contact voltage. Figure 10.6 explains the terms potential funnel and step voltage. The voltage drop results e.g. from lightning or, as shown in, if a wire of an overhead power line (e.g. because of a fault) is in contact with the ground. The earth is a poor conductor which is why there are differences in potential at different points of the Earth's surface. Standing on the ground, your feet tap different voltages, which can be life-threateningly high. Step voltage refers to the voltage that is tapped at a length of step of 1 meter.

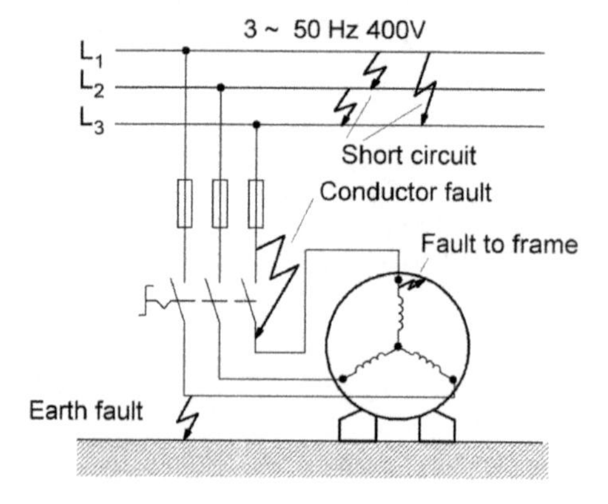

Figure 10.4: Different types of faults.

110 One countermeasure is called "spark suppression".

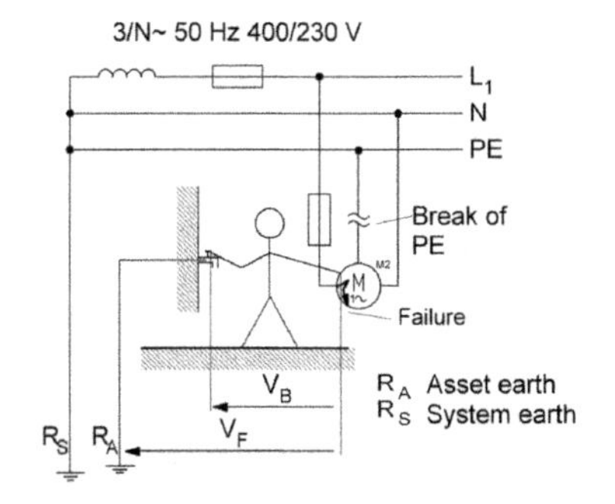

Figure 10.5: Fault voltage vs contact voltage.

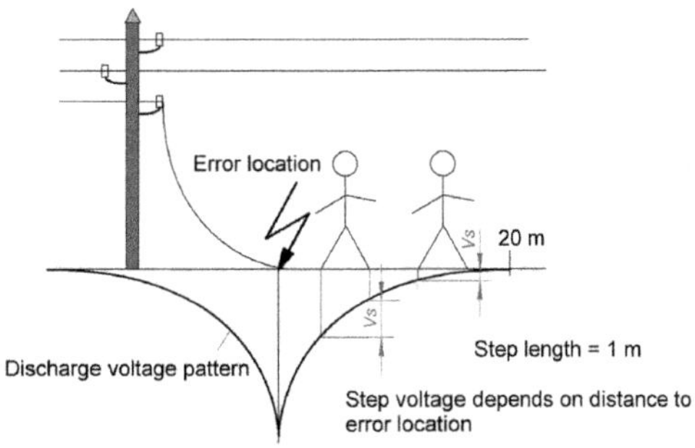

Figure 10.6: Potential funnel, step voltage (length of step: 1 m).

10.4 Review questions

1. Name several direct impacts on humans. Give examples.
2. Name the starting levels of voltages (AC, DC) that are fatal for humans.
3. The wire of an electric fence delivers electrical impulses > 4 kV to earth. Why does this not constitute a mortal danger to humans and animals?
4. Name fatal indirect effects of electricity on humans.
5. How does a residual-current device (RCD) work?
6. What level of current is needed for humans to be able to perceive alternating current?
7. What are the differences between the terms “short circuit”, “conductor fault”, “body fault” und “earth fault”?
8. What is a voltage drop? How can you keep the voltage tapped with your feet as low as possible?
9. Why do birds not suffer a fatal electrical accident when sitting on a wire of an overhead power line?

Bibliography

[1] Klaus Tkotz (ed.): „Fachkunde Elektrotechnik“, Europa Fachbuchreihe, Wien, 26. Version, 2009.

[2] Gert Hagmann: „Grundlagen der Elektrotechnik“, ISBN 978-3-89104-725-5, AULA 2009.

[3] Wolfgang Bieneck: „Elektro T, Grundlagen“, Holland+Josenhans Verlag, 7. Version, 2010.

[4] Heinrich Hübscher, Jürgen Klaue: „Elektrotechnik Grundbildung“, ISBN 3-14-231030-4, westermann 2000.

[5] Reinhold Pregla: „Grundlagen der Elektrotechnik“, ISBN 3-7785-2680-4, Hüthig 1998.

[6] G. Schmitz, Elektrotechnik: „Grundlagen der Elektrotechnik“, bookbon.com, ISBN 978-87-7681-786-2, eBooks 2011

[7] Hans Rudolf Ris: „Elektrotechnik für Praktiker“, ISBN 3-905214-71-7, electrosuisse Verlag, 2011.

[8] Carl H. Hamann: „Elektrochemie“, ISBN 3-527-31068-1, Wiley-VCH Verlag, 4. Version, 2005

[9] Peter W. Atkins: „Physikalische Chemie“, ISBN 3-527-25913-9, VCH Verlagsgesellschaft, 2. Version, 1990

[10] Giersch/Harthus/Vogelsang: „Elektrische Maschinen“, ISBN 3-519-26821-3, B:G. Teubner Stuttgart, 3. Version, 1991

[11] C. H. Hamann, w. Vielstich; Elektrochemie; Wiley-VCH Verlag, 4. Version, 2005

[12] Giulio Milazzo; Elektrochemie: Theoretische Grundlagen und Anwendungen; Springer Verlag; 1952

[13] Arnold Berliner, „Lehrbuch der Physik in elementarer Darstellung“, Springer Verlag Berlin Heidelberg, 5. Version, 1934; Reprint of the original: TP Verone Publishing, 14. 11.2016

[14] Korthauer, Reiner; Handbuch Lithium-Ionen-Batterien; Springer Verlag; 2013

[15] *https://www.tugraz.at/fileadmin/user_upload/Institute/ICTM/education/downloads/Skript_Lithium-Ionen-Batterien.pdf* access: October 2018

[16] *http://www2.pc.tu-clausthal.de/edu/05-CYCLOVOLTAMMETRIE.pdf* access: October 2018

[17] *https://www.physik.uni-augsburg.de/exp4/FP_A/material/FP27.pdf* access: October 2018

[18] Script: Fuel cells – Laboratory Course Technical Chemistry II; V. Hacker, Christoph Grimmer, Stephan Weinberger, TU Graz – Institute of Chemical Engineering and Environmental Technology

[19] C. Sumereder, Dielectric Investigations at Superconducting Insulation Systems, Doctoral Thesis 2003, TU Graz

https://doi.org/10.1515/9783110521115-011

Index

https://doi.org/10.1515/9783110521115-012